# CROSS THE MERSEY

# CROSS THE MERSEY

## THE TRANSFORMATION OF A RIVER

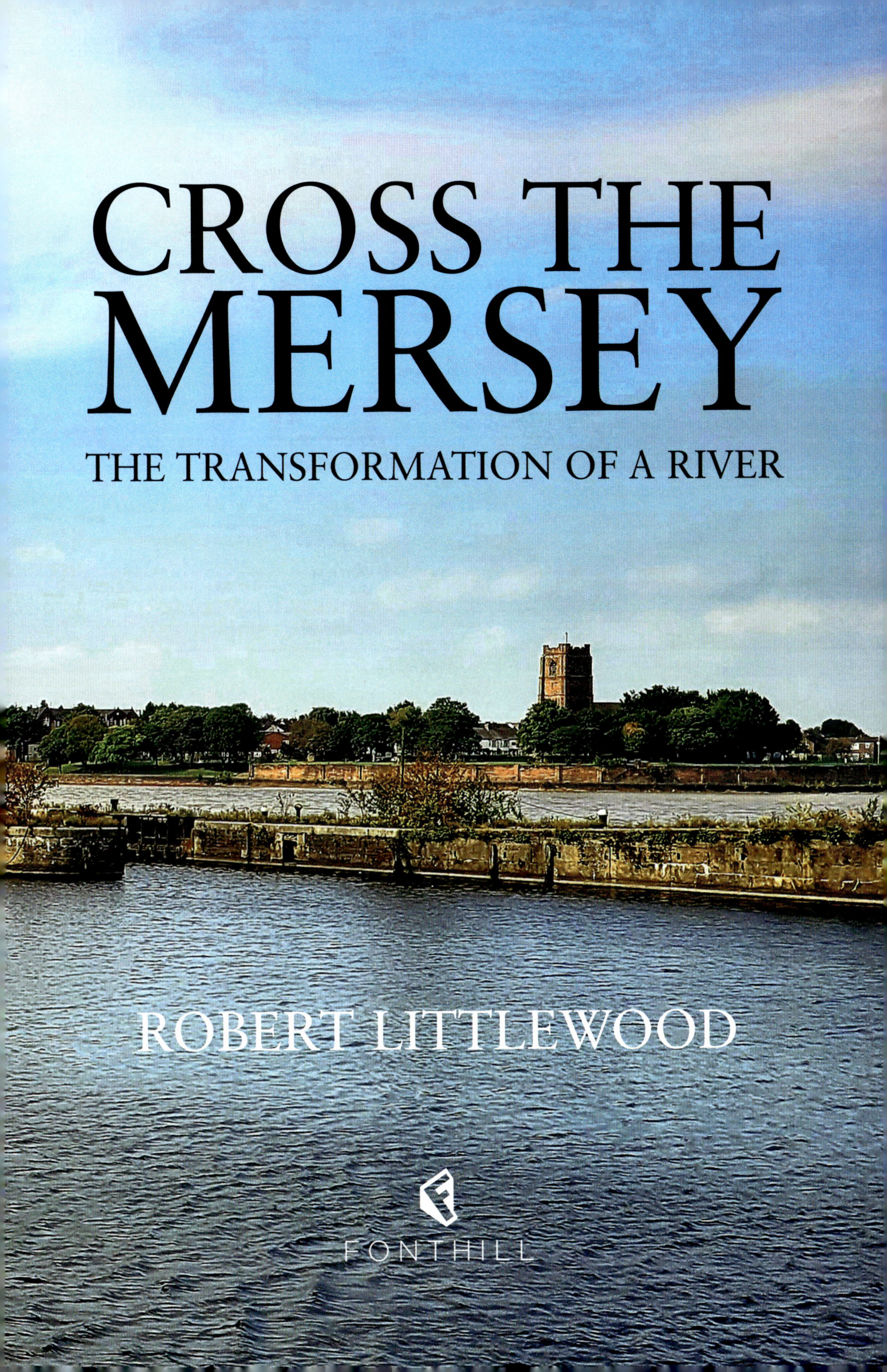

ROBERT LITTLEWOOD

FONTHILL

First published in Great Britain in 2026 by
Fonthill
An imprint of
Pen & Sword Books Ltd
Yorkshire – Philadelphia



ISBN 978-1-03614-737-2



A CIP catalogue record for this book is available from the British Library.



Typeset in SabonLTStd 10/13 by
SJmagic DESIGN SERVICES, India.
Printed and bound in India by Replika Press Pvt. Ltd.

MIX
Paper | Supporting responsible forestry
FSC
www.fsc.org
FSC™ C016779

The Publisher's authorised representative in the EU for product safety is Authorised Rep Compliance Ltd., Ground Floor, 71 Lower Baggot Street, Dublin D02 P593, Ireland.
www.arccompliance.com

For a complete list of Pen & Sword titles please contact
PEN & SWORD BOOKS LIMITED
George House, Units 12 & 13, Beevor Street, Off Pontefract Road,
Barnsley, South Yorkshire, S71 1HN, England
E-mail: enquiries@pen-and-sword.co.uk
Website: www.pen-and-sword.co.uk

or

PEN AND SWORD BOOKS
1950 Lawrence Rd, Havertown, PA 19083, USA
E-mail: uspen-and-sword@casematepublishers.com
Website: www.penandswordbooks.com

*To Joanne, Zoe, Eilise, Lily,*
*Dylan and Aliyah.*

# Acknowledgements

My thanks go to Stockport Heritage Trust, Warrington Borough Council, Chorlton Water Park and Bilway Marine Services.

My special thanks go to Steve Williamson for droning, checking and setting, plus invaluable help and support, and Joanne Williamson for maps, feedback and encouragement.

# Contents

Saddleworth Moor.

1

# The Mersey—A Brief History

From high on top of the Pennines, the River Tame winds its way down from the wild and windswept Saddleworth Moor, through Delph, Mossley, Tameside and Reddish to eventually meet the River Goyt in Stockport.

The River Goyt starts its journey on Axe Edge Moor in the Peak District, flowing north through Whalley Bridge, New Mills and Marple. Just beyond the 1813-built 'Iron Bridge' in Marple, the River Goyt will be joined by the River Etherow and continue to make its way to Stockport.

By an ancient ford in Stockport, the Rivers Tame and Goyt join and the Mersey embarks on its journey to the sea. The Mersey has over forty tributaries and confluences, each making their contribution to the river's seaward voyage.

'Mersey' comes from the Anglo-Saxon mǣres, 'of a boundary' and ēa, 'a river', and was once an ancient boundary between Mercia and Northumbria, later marking the boundary between Cheshire (south of the river) and Lancashire (north of the river). The river flows for approximately 70 miles from Stockport to the Irish Sea, although some still debate the location of the river's birth as an ancient map replaces the River Etherow with the Mersey, documenting its origin as the Derbyshire Moors, a further 26 miles upstream. A couple of later maps also record the start of the Mersey as being at the confluence of the River Goyt and River Etherow, but it is generally accepted today that Stockport marks the beginning of the river.

The Romans used the river and may have built a port as far upstream as Runcorn. A Roman road connected Chester (Deva Victrix) with Manchester (Mamucium) and would have crossed the Mersey near Sale, by ford or bridge. Ethelfleda, the eldest daughter of Alfred the Great, built a fortified settlement on the banks of the Mersey in Runcorn in AD 915.

The Vikings sailed the Mersey and named it the 'Mighty Serpent', establishing settlements around the Wirral and West Lancashire.

Fords and ferries provided crossing points across the river, close to developing rural communities. Benedictine monks from Birkenhead Priory, founded in 1150, operated a ferry across the Mersey, and the monks from Norton Priory ran their ferry at Fiddler's Ferry from Penketh to Moore on the south bank of the river from the early twelfth century.

Stockport and Liverpool both had castles and their proximity to the river would have been important. Nothing remains of either castle today but some of the stones

*Above left*: The River Tame by the Reddish Vale Viaduct.

*Above right*: The River Goyt at the packhorse bridge near Marple.

*Below*: Beneath the Wellington Road Bridge in Stockport looking towards the viaduct and Weir Mill.

from the Liverpool castle may have been repurposed by Thomas Steers when he built the world's first commercial wet dock in 1710.

The Thomas Steers Dock, known today as the Old Dock, almost bankrupted Liverpool but was a gamble well worth taking as it enabled Liverpool to thrive and become a major port. Industry grew around the port and further upstream in Manchester, Lancashire and Cheshire, where mills and factories manufactured cotton, silk and other textiles. Widnes became a centre of chemical manufacturing and Warrington would become synonymous with wire-making, brewing and tanning.

Manchester earned the nickname 'Cottonopolis', world-famous for revolutionising the cotton industry to become the international centre of textile manufacturing. The Port of Liverpool quickly developed to become one of the biggest ports in the UK, exporting goods around the world. From the 1750s the port would also dominate Britain's involvement in the slave trade, which continued even after slavery was abolished, a shameful period of our history.

The Runcorn Railway Bridge.

Liverpool Docks.

The Industrial Revolution brought with it the need for an increase in transportation and the Mersey and Irwell Navigation Act was passed in 1720. Many cuts and diversions were made in both waterways in an effort to reduce journey times and allow larger boats to access ports further upstream.

The oldest existing road bridge over the Mersey was built in Stockport in 1826, eleven years after Wellington defeated Napoleon at the Battle of Waterloo. The bridge was named Wellington Road Bridge in recognition of his achievements.

As industry grew along the river, so did the pollution and this would have a devastating effect on wildlife. The mills and chemical plants pumped toxins into the Mersey for decades, effectively poisoning the river as well as perverting its natural flow with cuts and locks, the worst of which would come in 1887 when the river would be butchered to form part of the Manchester Ship Canal.

The Liverpool and Manchester Railway, the world's first intercity railway, was completed in 1830 and revolutionised rail transportation. By the late 1800s, the Mersey and Irwell Navigation became heavily silted and ran into disrepair, which saw a decline in shipping—but that was by no means the end of the story for the transportation of goods via the waterways. Mill owners in Manchester sought a solution to bypass the docks at Liverpool as they felt charges had become excessive. This would eventually lead to the creation of the Manchester Ship Canal, built between 1887 and 1893, first used in January 1894.

The construction of the railways would see some impressive railway bridges span the Mersey, including the Warrington 'Twelve Arch Railway Bridge', built in 1837, the Runcorn 'Britannia Bridge', built in 1868 and the most impressive twenty-seven-arched Stockport Railway Viaduct, built in 1840, all still in use today.

For two centuries, the Mersey had endured man's most foul indignities. We had perverted its course, polluted its banks, and contaminated its waters. By the mid-twentieth century, the River Mersey was considered to be one of the most polluted rivers in Europe, and in 1985 the River Basin Campaign was established to improve water quality and encourage waterside regeneration.

From early in the twenty-first century, the Mersey has seen a dramatic transformation and flora and fauna have returned to its banks. Salmon can once again be seen in the river and even spawn upstream as far as the River Goyt. Sharks, dolphins and seals have been spotted around Liverpool Bay. By 2009, it was announced that the river was one of the cleanest in the UK, although reports still indicate high levels of micro-plastics.

The Mersey is considered sacred by British Hindus, who adopted the river and celebrate Ganesh Chaturthi, a ten-day festival marking the birthday of Lord Ganesha, the god of intelligence and wisdom. Hundreds of British Hindus attend an annual event where statues of the deity are submerged in the river to symbolise the beginning of his journey to the Himalayas, taking all humanity's adversities with him.

The river has lent its name to five ships, the latest being HMS *Mersey*, a River-class patrol vessel launched in 2003. The ship and crew attended a ceremony in Liverpool in April 2023, where she was awarded the freedom of Sefton. A major refit was completed at Falmouth and the ship returned to duty in July 2024.

The Mersey has arguably played a part in shaping Britain like no other river, lending its name to Merseyside, the musical genre Merseybeat, and the 1960s music paper *Mersey Beat*, created by Bill Harry, one of John Lennon's school friends. From Stockport to Liverpool, people have grown up by the side of the Mersey—and many more have felt its effect in literature, verse, or possibly the famous song by Gerry and the Pacemakers, *Ferry Cross the Mersey*, which is played endlessly on the ferries that still cross the Mersey today.

The River Mersey near Northenden.

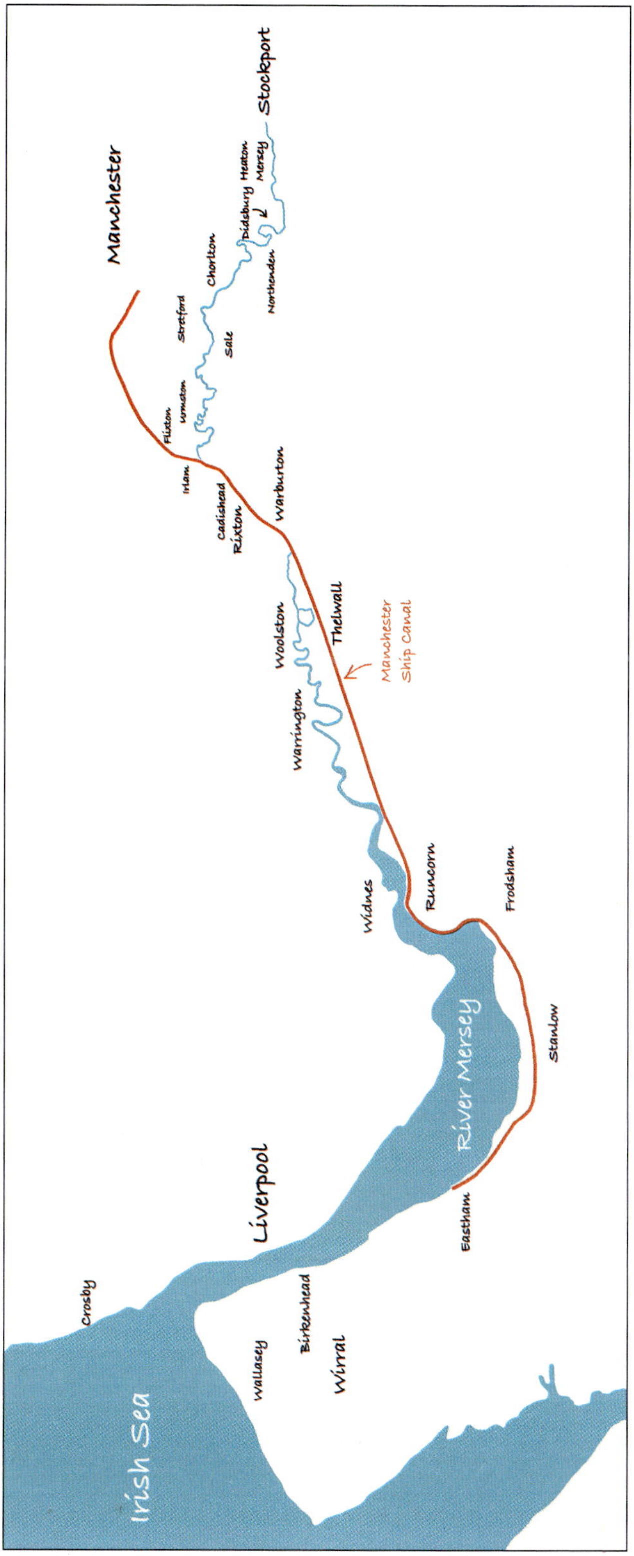


Map of Manchester Ship Canal and the River Mersey.

2

# My Mersey

I grew up in Heaton Mersey in Stockport, ten minutes from the River Mersey, at a point where Samuel Oldknow built his bleach works in 1785. We often played near the Mersey and even managed to get inside the Glazebrook to Woodley Line Railway Bridge, which proved to be surprisingly easy to access. You could climb through the bridge and out over the Mersey in the hollow wrought-iron sections. There was a thunderous roar when a train went over the bridge, which was utterly terrifying, even for the 'experienced explorer'. My parents never knew.

I remember many walks along the Mersey in the late 1970s and early 1980s. The river was toxic then and my father used to regularly warn us to 'never go in the water, don't even dip your toe in', which was a warning we definitely heeded. The river and banks were filthy, the industry along the banks was crumbling and left abandoned. We even came across a stolen car at one point, pushed or driven into the Mersey near the Cheadle Bridge. We did swim in water close to the Mersey—at the Galleon outdoor

Evening along the Mersey in Didsbury.

*Above*: Heaton Mersey Weir.

*Below*: The River Mersey in West Didsbury.

pool in Didsbury, Manchester's first lido, sadly now closed, but part of the Waterside Leisure Club.

I studied photography at college in Chorlton-cum-Hardy in the mid-1980s and often took photographs of the Mersey around Stockport, Didsbury and South Manchester, with the river already being of special interest to me. I remember my first trip to Liverpool and an altogether new and astonishing view of the Mersey and the landmark buildings that dominated the landscape. Previously only glimpsed on TV, the river and docks appeared eerie and foreboding, but I would soon learn more about the legacy of the Mersey and recognise its historical significance to the North West, Britain and the world.

Between the mid-1990s and early 2000s I worked in Stockport, at both ends of the shopping centre. I was the first manager of Dillons Bookstore on Merseyway and the last manager of Hammicks Bookshop on Warren Street. The river ran beneath us, through the Merseyway tunnel. I was also briefly a caretaker manager for the Dillons branch in Liverpool and the fated Hammicks Bookstore in Birkenhead—I never strayed too far from the Mersey.

Today, I live in Didsbury, five minutes from the Cheadle Bridge that crosses the Mersey in East Didsbury, the river still close to my heart and a constant in my journey through life.

While researching and photographing the river, I have discovered and fallen in love with Spike Island, Runcorn Promenade, Hale Lighthouse and Woolston, places I had previously passed without a second glance. Woolston would become something of an obsession, as would the many bridges and crossing points—there's not only a ferry that crosses the Mersey—and am amazed at how rare and architecturally unique some

The River Mersey at East Didsbury.

*Above*: The Mersey at West Didsbury.

*Right*: The start of the River Mersey in Stockport alongside the M60.

of the bridges are. I gave myself the challenge of capturing the beauty of the Mersey with the additional task of cataloguing all the bridges and tunnels that currently cross above or beneath its waters.

The River Mersey and its banks have transformed considerably since my childhood and I still find myself drawn inexplicably to the river of my youth. This book attempts to record a little of the river's amazing history and mystery, as well as document its beauty and remarkable transition from what was once described as an 'open sewer', back to a thriving, affluent waterway.

3

# Stockport

The River Goyt flows through Stockport, twisting between the remnants of old mills and the ruins of industry; after it passes under the stone-built Portwood Bridge it's joined by the River Tame. The rivers mingle and the Mersey is born.

Stockport is an industrial town in the North West of England, lying approximately 7 miles south of Manchester. The River Mersey once split the town in two, with the south bank of the Mersey being in Cheshire and the north bank being in Lancashire. The name is most likely derived from 'stoc', meaning hamlet, and 'port', meaning

*Below left*: Stockport Railway Viaduct.

*Below right*: Mersey Square in Stockport with the iconic Plaza to the left and Wellington Road to the right.

marketplace, although the stock could also refer to a stockade or castle; Stockport also once had a motte-and-bailey castle located close to where the town's historic market hall is sited today. Stockport developed exponentially during the Industrial Revolution and become renowned for hat making and silk spinning; the first water-powered silk mill in the North West of England was opened in Stockport in 1732.

It is perhaps not the most auspicious start for such a celebrated river, with the M60's concrete fortifications dwarfing the newly created waterway, but better things are to come—though not yet. The river flows

*Right*: St Mary's Church.

*Below left*: Knightsbridge.

*Below right*: The Mersey disappears beneath Merseyway.

*Above*: The River Tame (bottom left) passes beneath the M60 to meet the River Goyt (top right) beyond the weir.

*Below*: Lancashire Bridge.

under Knightsbridge, the plainest of bridges, and unceremoniously disappears under Merseyway, Stockport's remarkable shopping precinct. Remarkable not only because it was one of the first shopping precincts in the country, but also because it covers 435.7 metres (1,429 feet) of the river, making it arguably the widest bridge in the country.

The river makes a brief appearance at Lancashire Bridge, covered in the 1930s when the Merseyway Road was built over the river, and later the Merseyway Shopping Centre, which opened in 1965. The bridge was built in 1891 but uncovered in 2015 as part of a £1.6 million council project to restore the bridge to its former glory.

The Lancashire Bridge is an ancient crossing point of the Mersey, with records of previous bridges at this location dating back to 1282 (the earliest record of any bridge crossing the Mersey) but there is written evidence of an earlier ford close to this location too, which just might have been used by the inhabitants of Stockport's long-demolished castle.

The Mersey resurfaces at the 'Bear Pit' in Mersey Square, where rumour has it that bears were kept to taunt and bait, possibly from as early as the twelfth century. A cave close to the Stockport Air Raid Shelters museum on Chestergate is called 'Bear Hole', but there is no real evidence that bears were kept here and it seems unlikely they were ever goaded at this location as it was only created in the 1930s, and would have been open to the river until that date.

Immediately south of Mersey Square is Stockport's 'other' viaduct, Wellington Road Bridge, the oldest existing road bridge to cross the Mersey. Built between 1824 and 1826,

*Below left*: Beneath Merseyway, lit for maintenance in the evening.

*Below right*: Stockport's indoor market (right).

Wellington Road Bridge with the brick staircase leading from Wellington Road and made famous by artist L. S. Lowry who was photographed at the top of the stairs sketching the view in 1961.

the eleven arches of the bridge (not all visible today) span the Mersey and the lower river valley through Stockport. The bridge is Grade II listed, designed by James Nowell, and named after the Duke of Wellington, who defeated Napoleon at Waterloo in 1815. The width of the arch over the Mersey is an impressive 27.4 metres (90 feet), an amazing achievement for its time, though it sadly collapsed on its first attempt, killing two workers.

Before the railway viaduct, the river flows under the new Astley Street Bridge, constructed in 2018—part of Stockport's new transport interchange, which combines apartments, an impressive town centre park and a spectacular spiral ramp with views over the Mersey. On the opposite side of the new transport hub, you will find Hat Works, Stockport's fascinating hat museum, which showcases much more than just the town's history of manufacturing headwear.

Stockport's landmark railway viaduct was built in 1839–1840 by the architect John Lowe, in partnership with engineer George Watson Buck. The initial two-track, brick-built, twenty-seven-arch structure was constructed using 11 million bricks; however, the viaduct was widened to four tracks in 1887–1889, taking the total brick count used to around 22 million. It remains one of the largest brick-built structures in the UK and is still an impressive sight today, almost 200 years after its construction.

*Above*: Wellington Road Bridge with the River Mersey exiting Merseyway beneath.

*Right*: Mersey Square with two arches of the Wellington Road Bridge visible behind the 'Bear Pit'.

*Above*: Astley Road Bridge in front of Stockport Viaduct.

*Below*: Stockport Interchange.

*Right and below*: Stockport Interchange.

King Street West Bridge with Stockport Viaduct and Weir Mill behind.

Standing partially under one of the arches of the railway viaduct, but built before it, is Weir Mill (originally Wear Mill), built in 1790 by John Collier but sold to Thomas Fernley in 1824. Fernley partially rebuilt, improved and expanded the mill after a devastating fire in 1831. The mill predominantly used spinning mules, invented by Samuel Crompton between 1775 and 1779, used to spin textile fibres into thread. (Working spinning mules can still be seen at Quarry Bank Mill in Styal.)

Built alongside Weir Mill is the magnificent stone triple-arch King Street West Bridge, constructed between 1850 and 1860. A footpath is accessible along the north bank of the Mersey from this bridge all the way to the Cheadle Bridge.

On to the Hollywood Way Bridge, which opened in 1992 and carries six lanes of the A5145 over the Mersey, replacing the two lanes of the Brinksway Bridge. Now pedestrianised, the stone-built single arch of the Brinksway Bridge lies close to the Hollywood Way Bridge but dates back to the late 1700s. It's a beautiful bridge, originally built for mill workers, and was widened and strengthened in 1847, the date prominently displayed on the keystone of the northeastern-facing arch.

The river passes over another weir and once again there is evidence of old mills and factories in abundance. On the south side of the river are the Brinksway Caves, once linked to a network of caves under Stockport and were expanded and used as air raid shelters in the Second World War. Originally thought to have been built by mill workers and expanded by navvies who lived in them during the construction of the Stockport Railway Viaduct, the caves are also known as 'Maggie's Cave', after a local witch that used to frequent them.

*Above left*: King Street West and Weir Mill.

*Above right*: Stockport Viaduct and King Street West.

*Below*: Brinksway Bridge from beneath the Hollywood Way Bridge.

*Above*: Brinksway Weir.

*Left*: Hollywood Way.

*Right*: Brinksway Caves.

*Below*: Geoff Funnell Bridge.

The Geoff Funnell Bridge was constructed in 2006 and replaces an earlier bridge, the Gorsey Bank Bridge, which had been at this location since 1894 but had fallen into disrepair and become unsafe. A plaque close to the bridge reads: 'In Memory of Geoff Funnell and Flann Griffin of Stockport Metropolitan Borough Council in recognition of their work in maintaining and improving the public right of way network in Stockport.'

The Mersey and the M60 ring road maintain a close relationship throughout South Manchester and the two rarely stray too far from each other, but here at Heaton Mersey it is the first time the river passes underneath the road at this functional but very plain 1982-built motorway bridge. A short distance along the path, you will notice more evidence of our industrious past; two railway bridges once crossed the river here: the Chorley to New Mills (Midland Railways) bridge, and my childhood favourite, the Glazebrook to Woodley (Cheshire Line) bridge built in 1865—this line once continued in a westerly direction to the Cadishead Viaduct and over the Manchester Ship Canal.

A further 300 yards (274 metres) along the path, the river flows over another weir, alongside which once stood Samuel Oldknow's Heaton Mersey Lower Bleach Works, built in 1785. Close to the weir, you may notice the remains of a sluice gate that allowed water to flow into the mill to power a water wheel. A goit—a small water channel—used to exist, running from the mill and back into the Mersey further along the footpath, where a disintegrating bridge supporting a water pipe crosses the Mersey adjacent to a sewage plant on the south bank of the river.

*Below left*: M60 motorway bridge over the Mersey.

*Below right*: The remains of Heaton Mersey Bleach Works with the sluice gate just visible on the far right.

4

# Didsbury–Northenden

Just under a mile along the riverbank and you will reach the stone-built three arches of the picturesque Cheadle Bridge, built in 1861. There have been at least three bridges built at this location: the earliest recorded being in 1755, which was destroyed in floods in 1756; the bridge was subsequently rebuilt in 1758, 1780 and finally in 1861, which is the bridge we see today. Although no recorded evidence has been uncovered, some speculate an early wooden bridge was located at this site.

Cheadle Bridge.

*Above*: Cheadle Bridge.

*Below*: The River Mersey alongside the Waterside Hotel and Leisure Club (left) and Micker Brook on the right.

Styal Line Railway Bridge with the Kingsway dual carriageway bridge seen through the arch.

At this point, the path comes to an end, but you can take a five-minute detour down the side of the Waterside leisure centre and hotel, which was the location of the Galleon outdoor pool. The pool originally dates back to the 1930s and is considered to be the first outdoor pool in Manchester. Although the hotel and gym now occupy the site the pool remains, although it is now indoors. In the mid-2010s, the original children's paddling pool was discovered precariously hidden under the weight-lifting room.

Back on the riverside path and we join the first of three 'Kingsway' crossings of the Mersey. This is the Kingsway dual carriageway road bridge built in 1930 to connect Levenshulme to Cheadle. Just a stone's throw away is the grand Manchester to Styal Line Railway Bridge, built in 1909. This magnificent black-brick, twelve-arched viaduct crosses the Mersey and the old Glazebrook to Woodley railway line.

For the next 10 miles, the riverbank is transformed into the first line of a flood defence and takes on a slightly artificial look with a two-tiered built-up bank and overflow catchment areas. The river meanders through leafy suburbs of South Manchester and walkers can enjoy the Mersey from both sides of the river.

Simon's Bridge in Didsbury is a small footbridge that crosses the Mersey near Ford Lane. Henry Simon (Gustav Heinrich Simon), a Prussian-born, Swiss-qualified engineer, moved to Manchester in 1860 and although penniless, he established himself as an engineer and revolutionised the flour industry with his 'rolling flour

*Above left*: Styal Line Railway Bridge.

*Above right*: Eight of the twelve arches of the Styal Line Railway Bridge.

*Below*: Simon's Bridge.

*Right*: Simon's Bridge looking across to Didsbury Golf Club and the continuation of Ford Lane to Northenden.

*Below*: A cold morning at Simon's Bridge in Didsbury.

milling' plant. In order to improve education for his daughters, Simon co-founded the Withington Girls' School and also, with his good friend Charles Hallé, was a founding member of the Hallé Concerts Society. His eldest son, Ernest Henry, would become the 1st Baron Simon of Wythenshawe. Henry Simon died in 1899 but left money for a bridge to be built at the location of a ford across the Mersey from Ford Lane to Poor's Field (now Didsbury Golf Club). Construction was completed in 1901 and, although quite possibly the smallest bridge to cross the Mersey, local Mancunians are terribly protective and passionate about their little River Mersey bridge.

The river continues alongside Withington Golf Club before passing under the M60 (built in 1982 as part of the M63) and towards Northenden.

At the site of the current footbridge, there was a ferry crossing dating back to 1641. An earlier toll bridge also spanned the river at this location, predating the current footbridge by at least a hundred years. On the south side of the river stands the Tatton Arms, designed by James Redford and originally named the Boat House Inn. The building has been derelict for a number of years but is due to be redeveloped.

Alongside the weir, at the site of the current park and conservation area, was a flour mill. A team of professionals from Manchester University and Manchester Museum, together with volunteers and local school children unearthed the remains of the mill, which, although rebuilt several times, stood on this site for over 700 years.

*Opposite*: Northenden footbridge with the M60 motorway bridge beyond.

*Below*: The Northenden Weir with the M60 motorway bridge and Northenden Bridge in the distance.

*Above*: Northenden Weir with the fish ladder visible on the opposite side of the river and the Tatton Arms to the right.

*Below*: Approaching Princess Parkway.

*Above left and above right*: Princess Parkway flanked either side by motorway slip roads.

A fish ladder was added to the weir in 2003 to enable salmon and other fish to pass further upstream. Salmon now spawn on a regular basis in the River Goyt.

The Northenden Bridge carries Palatine Road across the Mersey. There has been a road bridge at this location since 1862, but the Mersey notoriously floods here and the bridge has been destroyed and replaced many times. The current bridge uses the abutments from much earlier bridges, although the piers are relatively modern (probably 1970s).

The river continues around three sides of Northenden Golf Course before it passes under Princess Parkway (A5103) and two slip roads to the M60, making three bridges in all. The Princess Parkway Bridge was built in 1931 and has some decorative features, but is somewhat hidden on both sides by the motorway slip roads and some very garish graffiti.

5

# Chorlton–Sale–Stretford

Round the next bend in the river, on the south side is Kenworthy Wood, a secluded knot of woodland which includes an orchard with apple, pear and plum trees, and on the north side is Chorlton Water Park. Chorlton and its neighbouring Sale Water Park were once small quarries used to excavate aggregate for construction of the M63 (now M60) in the early 1970s. When the motorway was complete, both quarries were flooded to create the two conservation areas and water parks.

Chorlton Water Park has its own river crossing in the form of a Bailey bridge. The original bridge, initially used to transport waste materials to the now-disused Barlow Tip, was erected in the early 1970s but became unsafe and was removed in early 2025, with a replacement due later in the year.

Barlow Hall Refuse Bailey Bridge—known locally as the Bailey Bridge (removed in 2025).

The Bailey bridge was invented by Donald Bailey for the British military in the early 1940s for use as a portable, temporary, quick-to-construct crossing point of a river or obstacle, or where a bridge had been damaged or destroyed.

The South Manchester Electricity Substation overlooks the south side of the Mersey and a small, haunting, concrete cable bridge, not for public use, crosses the river.

Around the next bend is a small Metrolink bridge built in 2012, connecting Manchester Victoria to Manchester Airport; visible just beyond that is the famous Jackson's Bridge.

South Manchester Electricity Substation Cable Bridge (not accessible).

*Above*: South Manchester Electricity Substation, locally known as Fairy Lane Substation.

*Below*: Jackson's Bridge with the Metrolink Bridge in the distance.

Jackson's Boat.

Jackson's Bridge.

Arguably, the two most famous bridges in Greater Manchester are the smallest, both prized by locals and steeped in history. We've crossed Simon's Bridge in Didsbury and can now cross Jackson's Bridge, linking Chorlton to Sale.

This incarnation of Jackson's Bridge was built in 1881 and takes its name from a local farmer, who used to ferry people across the Mersey at this location for a small fee. The nearby pub, Jackson's Boat, also takes its name from the farmer and his ferry, although previously it had been named Bridge Inn. When the bridge was damaged in 1880, the publican John Brookes commissioned the Edward Taylor Bellhouse Foundry to construct a new wrought iron toll footbridge which was completed in 1881. The toll fee was eventually abolished in the late 1940s.

The river weaves its way around Sale Water Park, Stretford Weir and under another Metrolink bridge, formally a railway bridge built in 1849 to carry the Manchester London Road to Altrincham (Great Central Railway LNWR) line. Half the bridge was converted to carry trams to Altrincham in 1992.

The ancient Barfoot Aqueduct can be seen running parallel to the railway bridge and carries the Bridgewater Canal over the Mersey. Constructed in 1765 by James Brindley, the bridge was commissioned by Francis Egerton, the 3rd Duke of Bridgewater, and is the oldest existing bridge to cross the River Mersey. The original stretch of the canal was opened in 1761 in order to transport coal from Egerton's mines in Worsley to Manchester but was later extended to Runcorn, with the first part of the extension opening in 1767.

*Above*: Barfoot Aqueduct seen from beneath the former Great Central Railway line bridge.

*Below left*: The former Great Central Railway Bridge (Manchester–Altrincham) now used as part of the Metrolink to carry trams.

*Below right*: The Barfoot Aqueduct carries the Bridgewater Canal over the River Mersey.

The footpath disappears at this point but walkers can take the path bearing right, along the tram line, and join the Bridgewater Canal towpath, heading in a southerly direction. Continue over the Barfoot Aqueduct and under the M60, take an additional path across Crossford Playing Fields and rejoin the Mersey at Crossford Bridge.

Meanwhile, the Mersey skirts around an animal sanctuary and under the M60 with the two slip roads of junction 7 either side. Here the Mersey leaves Manchester's ring road for good and heads in a westerly direction towards Crossford Bridge, Sale and Urmston.

Crossford Bridge is another beautiful, single-arch, stone-built bridge thought to have been built in the 1850s, though an exact date has not been recorded. Earlier records reveal that there has been a bridge at this location since 1538, although as this was the route of the Roman road linking Chester to Manchester, it might have been the location of an even earlier bridge or ford. The 1538 bridge was damaged in floods and rebuilt in 1578, only to be totally destroyed in 1745 by Bonnie Prince Charlie to slow the advance of the Jacobites. The current bridge was widened in 1907 with additional work being completed in 1961 when the neighbouring steel bridge was built to carry south-west-bound traffic on the A56.

The north bank footpath resumes from Crossford Bridge, and the Mersey circumvents another golf course (Ashton on Mersey) before passing under the Ashton Bridge, built in 1987 to carry the Carrington Spur (A6144) over the river along with its neighbour, the Stretford Millennium Footbridge. The Millennium Footbridge spans the Ashton Weir and links the picturesque Ashton-on-Mersey to Cob Kiln Lane, Urmston Cemetery and Urmston Meadows.

Crossford Bridge.

*Above left*: The new Crossford Bridge with the old Crossford Bridge just visible beneath.

*Above right*: The River Mersey flows beneath the M60 motorway for the final time, leaving Stretford and heading towards Sale.

*Below*: Ashton Bridge with the Stretford Millennium Footbridge just visible beneath.

*Above left*: The Millennium Bridge spans the Ashton Weir.

*Above right*: Millennium Bridge linking Ashton-on-Mersey with Cob Kiln Lane.

*Right*: The River Mersey through Urmston.

6

# Urmston–Flixton–Cadishead

The river is at its most untamed here and flows riotous and unbound through Urmston and into Flixton. I love this stretch of the Mersey; it's wild and unkempt. There is something new around every corner and you will often find eroded embankments or flotsam deposited along the shore. Sadly, land on both sides of the Mersey soon becomes privately owned and we must take a detour. As the river flows by another golf course (Flixton), take Church Road (B5213) and Carrington Road (B5158) to Flixton Bridge, a single-arch, stone-built bridge constructed in 1907. A pipe crosses the river alongside the bridge, presumably for industrial purposes, as the Mersey will soon be flanked by Carrington Power Station on its south banks and the vaguely visible remains of sewage works on the north banks.

If you follow Carrington Road to its conclusion, a small footpath will lead you to the little-known Wellacre Country Park, once the site of a sewage works; evidence of the demolished plant can be seen in the fields and bordering the river. The Mersey

*Right*: Flixton Bridge.

*Below*: The River Mersey flowing through Urmston and on towards Flixton.

Aerial view of the Mersey meandering through Urmston.

Two bridges carrying pipes cross the Mersey to the long demolished sewage works with Carrington electricity substation (top left) and haulage company on the south bank of the river.

is visible from the park and four further aqueduct and cable bridges, presumably no longer used and definitely not open for public access, cross the river along this 2-mile stretch. The park runs on both sides of the Manchester to Liverpool railway line with two tunnels passing beneath the railway line, enabling pedestrians to access both sides. The railway crosses the Manchester Ship Canal on the striking Irlam Railway Viaduct which is next to the Mersey Weir, the point where the River Mersey joins the ship canal—the river and the ship canal are now one, at least for the next 4 miles.

The Manchester Ship Canal can be crossed on foot at Irlam Locks, and pedestrians can view the Mersey entering the ship canal a short walk along the A57, just beyond the Irlam Railway Viaduct.

The Irlam Railway Viaduct was opened to freight traffic in January 1893 and passenger traffic in March 1893. The bridge originally consisted of brick-built arches on either side of the 37 metres (121.3 feet) central steel span, but these were replaced by further steelworks in October 1893 when cracks appeared in the brickwork. In the thirteenth century Irlam was known as Irwellham, as it sat either side of the River Irwell and was largely undeveloped until the Irlam Railway Station opened in 1873.

*Above left*: The Mersey just before it enters the Manchester Ship Canal with the Manchester city skyline in the distance.

*Above right*: The Mersey with Carrington on the south bank of the river. Note (very top left corner)—the Manchester Ship Canal is just visible as it passes beneath Cadishead Railway Viaduct.

*Below*: The Mersey as it enters the Manchester Ship Canal alongside the Irlam Railway Viaduct.

*Above left*: Irlam Locks with a view of Irlam Railway Viaduct.

*Above right*: Irlam Railway Viaduct.

*Below*: *Daniel Adamson* 'The Danny' on the River Mersey.

The Manchester Ship Canal used a large portion of the Mersey and Irwell Navigation, approximately 4 miles of the River Mersey and nearly 8 miles of the River Irwell, and in turn this was widened and deepened to create the Manchester Ship Canal.

A few meanders and off-shoots of both rivers can still be found; most notably, the Irwell has a fork opposite Hulme Ferry and an even larger strip at Irlam, near the Boat House Inn. Remnants of the Mersey's original course are visible around Warburton and Lymm.

Engineer and foundry owner Daniel Adamson was the driving force behind the creation of the Manchester Ship Canal. Fellow engineer Edward Leader Williams submitted plans to dig a 36-mile-long canal with five locks, at an average depth of 28 feet (8.5 metres) from the Port of Manchester to the estuary on the Wirral coastline. Construction began in 1887 and was completed in December 1893, opening to traffic on 1 January 1894. Daniel passed away in January 1890 so would never see the canal completed, but a refurbished tugboat, commonly known as 'The Danny', was named after him. Built by Cammell Laird shipyard in Birkenhead in 1903, the little tug was extensively refitted in 1936 to transport gentry along the ship canal and is still in use today conveying passengers on sightseeing trips along the canal.

First the river, now part of the Manchester Ship Canal, must pass under the derelict Cadishead Railway Viaduct, still an imposing sight. It opened in 1893 by CLC (Cheshire Lines Committee) to carry the Glazebrook to Woodley Line, and was constructed of brick and steel lattice girders. The viaduct was closed in 1983 when it became evident that extensive and expensive repairs were required. Shipping containers

The Manchester Ship Canal near Lymm.

*Above*: Irlam Locks.

*Below*: Cadishead Railway Viaduct; notice the shipping containers placed either side to stop people 'exploring' the bridge.

*Above left*: Irlam Railway Station.

*Above right*: View from the top of Cadishead Viaduct.

now block each end of the viaduct as the bridge was being used as a battleground between warring teenagers from Cadishead and Partington. Hamilton Davies Trust, a local community charity, aim to restore the viaduct as part of a heritage railway project. The charity refurbished the derelict station at Irlam and developed a splendid restaurant and bar with heritage railway displays.

A short distance from the Cadishead Viaduct, on the north bank of the canal, a tiny footpath meets the A57. The footpath is Bob's Lane and lent its name to a ferry that carried paying passengers across the Manchester Ship Canal, commencing on New Year's Day 1894, the day the canal officially opened. A similar ferry had operated at this point when this section of the canal was still the River Mersey.

On Tuesday 14 April 1970, at 5.30 pm , several explosions shook local buildings and a mile-long stretch of the ship canal exploded into a wall of flames, 60 feet (18 metres) high, tragically engulfing the ferryman Bernard Carroll and his five passengers. The fire was caused when gallons of fuel overflowed into the canal near the Shell chemical plant, where negligent workers were refuelling a tanker at the Partington Coal Basin. The ferry reopened shortly after the disaster but the business never recovered and the ferry finally closed for good on 18 January 1993. An information and memorial plaque was erected where Bob's Lane meets the A57 in Cadishead.

A memorial cruise was held on Wednesday, 30 April 2014, from Salford Quays down the ship canal to the site of the disaster where family and friends placed flowers and wreaths.

*Above*: The Manchester Ship Canal with Cadishead on the left and Partington on the right near the location of Bob's Lane Ferry.

*Below*: The Hulme Ferry, one of two ferries still in operation today run by Bilway Marine Ltd.

Information and memorial plaque for Bob's Ferry.

7

# Rixton–Warburton

Along the A57, at the roundabout where Cadishead Way, Liverpool Road and Manchester Road meet, a small footpath running along the north bank of the ship canal commences and runs to the Rixton and Warburton Bridge, more commonly known as the Warburton Toll Bridge. Sir Edward Leader Williams, designer of the Manchester Ship Canal also engineered the Warburton Toll Bridge as well as its almost identical sister bridge, the Ackers Road Bridge.

Following an Act of Parliament in 1863, the original Rixton and Warburton Bridge was built in 1864 over the Mersey, which ran through the village of Warburton. When the Manchester Ship Canal was built in 1893, the river was diverted from its original

Warburton Toll Bridge.

course through Warburton and, over the subsequent years, that section of riverbed dried up and was filled in. Along with the ship canal in 1893 came the 'new' cantilever Rixton and Warburton Toll Bridge and, due to the 1863 Act, only two shillings and sixpence (12½p) could be charged per crossing, up to a maximum of 25p per day. After 158 years of the 12½p toll charge and huge local opposition, the government agreed on an increase—and in autumn 2024 the price was increased to £1.00 and work commenced on repairing the bridge and approach road. The tollbooth was also removed and a modern vehicle registration scanning system will be installed with customers required to pay online. The stone abutments and railings are still visible from the original 1864 bridge close to the toll bridge cottage on the approach road to the cantilever bridge and a tiny stream can still be seen flowing under the bridge at various times of the year. The River Mersey? I like to think so.

Just over a mile west of the Warburton Toll Bridge, just as the River Bollin enters the Manchester Ship Canal, as if taking offence, the Mersey leaves and continues on its own journey to the Irish Sea. To view the Mersey leaving the ship canal is difficult as the land all around this area is private, but it is possible from the river itself, as is viewing our next bridge, the Rixton Railway Bridge, part of the huge Manchester Ship Canal rail network.

Originally of timber construction, the bridge was replaced in the 1920s with a reinforced concrete bridge, a similar example of which remains at Moore Nature Reserve near Port Warrington. The line ran along the north bank of the ship canal,

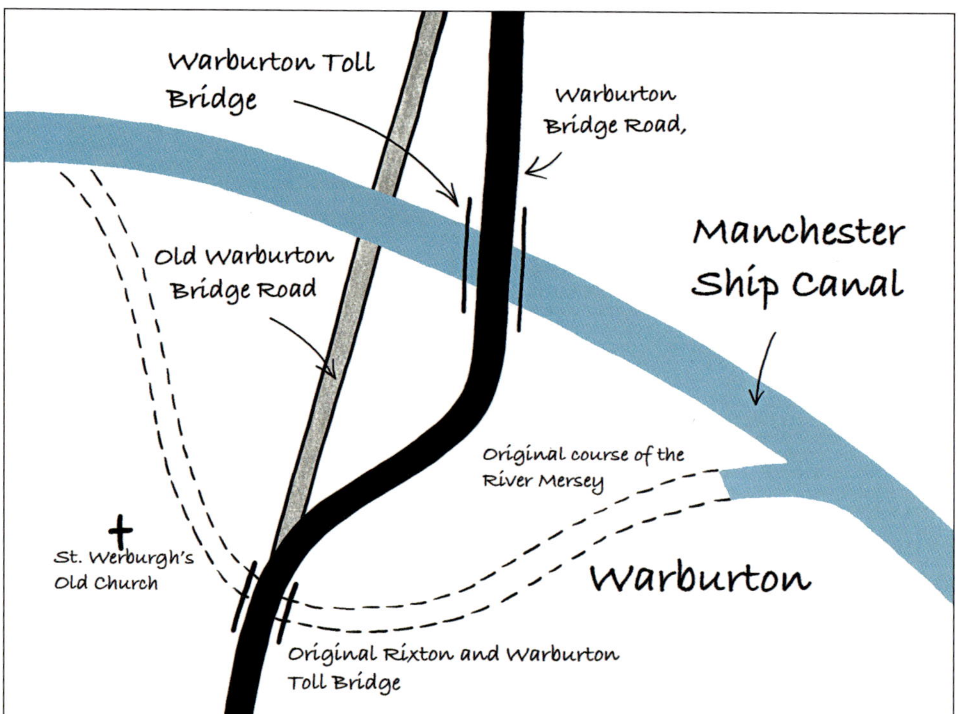


Map showing Warburton Toll Bridge.

*Right*: The original Rixton and Warburton Toll Bridge with stone abutments and iron railing still visible not far from the tollbooth.

*Below*: The Mersey resumes its original course heading north beneath the Manchester Ship Canal Rixton Railway Bridge and then west towards Woolston.

*Above left*: Another old Manchester Ship Canal railway bridge at Moore Nature Reserve …

*Above right*: … and opposite Hulme Ferry in Eccles.

*Below*: An original section of the River Mersey in Lymm still in water where several concrete barges were abandoned.

initially to transport materials during its construction but later to transport chemicals and other goods. The Manchester Ship Canal Railway Line was officially closed between Partington and Irlam in 1978.

Back on the north side of the canal, our path once again runs into private land and we must take a detour back along the A57 for 1.3 miles (2.1 kilometres) and join the start of the Merseyway footpath which will take us along a five-mile route into the heart of Warrington.

Where the Merseyway footpath meets the Mersey there was once a large horseshoe loop in the river. In 1829, a 330-metre (360-yard) cut was made in the river, effectively eliminating the loop in an attempt to make the river more navigable. This cut was called the Butchersfield Canal. The lowest part of the loop now lies south of the Manchester Ship Canal in Lymm Golf Club and still holds some water, as well as six ferrocrete (concrete) barges and a wooden vessel. Three of the barges were abandoned in the 1960s and are barely visible today, with a further three being discarded in the 1980s. The wooden vessel is believed to be *The Elk*, a frigate built in 1851 that served in the Crimean War (1854–56) and joined the Manchester Ship Canal fleet in 1890. In an attempt to trial a new method of dredging, *The Elk* was fitted with a suction pump and although not entirely successful at the time, it would become the preferred method for maintaining waterways and deepening ports. *The Elk* was involved in a collision with the SS *Hector*, but was repaired at Ellesmere Port in the early twentieth century before retiring in 1931 and being abandoned in this lost stretch of the River Mersey. A sad and unfitting end to a beautiful, historic little ship.

*Below left*: A concrete barge in an old section of the Mersey.

*Below right*: A concrete barge at the National Waterways Museum.

8

# Thelwall–Woolston

Just under a mile along the Merseyway path, you will reach the imposing Thelwall Viaduct, although you will probably hear the traffic long before you see it. There are actually two Thelwall Viaducts, the northbound bridge built in 1963, and the southbound one built in 1996. Before the viaducts is another Bailey bridge, used to carry landfill across the Mersey to bed 1 of Woolston Eyes Nature Reserve—'eyes' comes from the Anglo-Saxon 'ees', which means the land around a loop in a river.

Designated a Site of Special Scientific Interest, the Woolston Eyes Nature Reserve comprises four raised beds originally created from dumping dredgings from the Manchester Ship Canal. Beds 3 and 4 are now a sanctuary for wildlife and a bird spotters' paradise. It is open to the public by permit only. Check the Woolston Eyes website for up-to-date permit details.

*Above*: The two Thelwall motorway viaducts—side-by-side.

*Below*: A Bailey Bridge (left) alongside the Thelwall Viaducts.

*Opposite*: The Woolston Eyes Aqueduct was used to pump dredging from the Manchester Ship Canal.

*Left*: The River Mersey as it approaches the Thelwall Viaducts.

*Below*: Woolston Eyes.

Just over half a mile from the Thelwall Viaduct is the Woolston Guard Bridge, built in the 1940s to manage water levels further upstream and in the Manchester Ship Canal. Adjoining the Guard Bridge is a timber footbridge that leads around Woolston Eyes (beds 3 and 4), and as it joins the north bank of the ship canal, you will find the Woolston Footbridge, which crosses the loop in the Mersey and gives pedestrian access to bed 3

Woolston Eyes Guide Bridge.

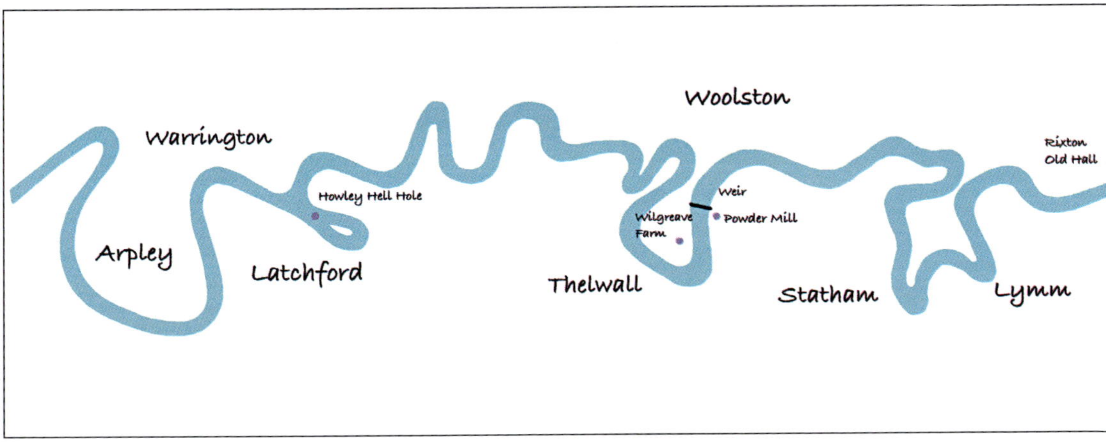


*Above*: The Mersey meandering through Woolston and Warrington around 1700.

*Below*: Map showing the canals and cuts between Arpley and Rixton Old Hall.

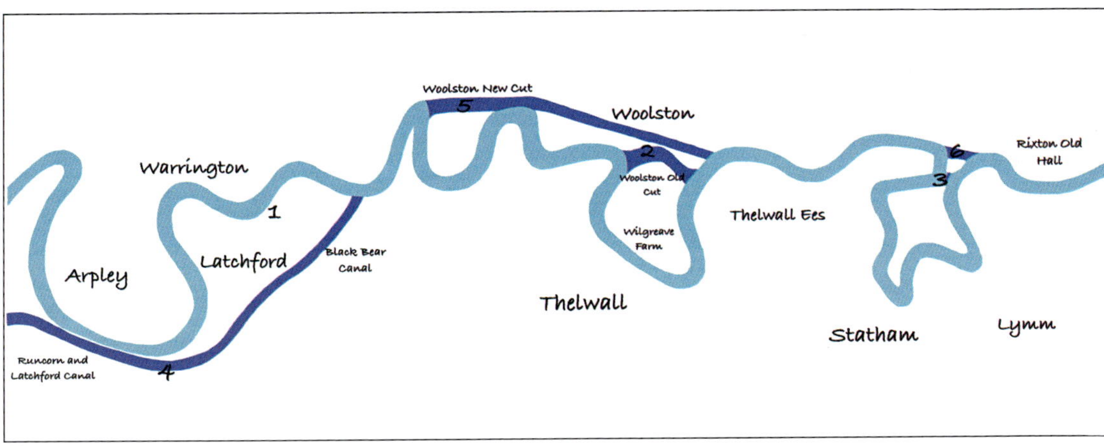


of the nature reserve. The bridge is a magnificent example of civil engineering, built on cast-iron cylindrical supports with wrought-iron trestles, and is thought to be the eldest cable-stayed bridge in England. Along with the Warrington Transporter Bridge, the Woolston Footbridge was nominated for a civil engineering heritage award in 2016.

## The Mersey and Irwell Navigation

Although local businessmen made proposals to make the River Mersey more navigable as early as 1677, it wasn't until 1712 when Thomas Steers, the designer of the Old Dock in Liverpool, surveyed land surrounding the Mersey and Irwell rivers and made a proposal to build eight locks and deviations to cut out meanders in the Mersey and Irwell to make the rivers more navigable and greatly reduce the time for shipping to reach Manchester.

The Mersey and Irwell Navigation Act was passed in 1721 with sufficient cash raised by 1724 to start work on the eight locks proposed by Steers, which were completed by 1734. Once navigable, additional work began to reduce travel time and bypass the natural meanders and horseshoe bends in the river around Woolston, Lymm and Warrington.

The major deviations and alterations made to the Mersey were:

1720s—A lock was built at Howley and the treacherous 'hell hole' bend in the river bypassed.

1755—A cut was made to the top of a large meander in the river which was the location of Wilgreave Farm and birthplace of 'Old Billy' in 1760. This cut would become known as 'Woolston Old Cut'.

1760—A lock was built between two bends in the river which would become known as the 'Old Lock'.

1804—The Runcorn to Latchford Canal was completed and connected Runcorn to the Mersey and Irwell Navigation at Warrington, bypassing some shallow sections of the Mersey near Fiddler's Ferry.

1821—A new cut was made in Woolston to bypass more bends in the river, with locks at Paddington Meadows and Woolston Weir. The new cut would effectively make the old cut redundant.

1829—A 360-yard (329-metre) stretch known as the Butchersfield Canal was cut to avoid a meander around Lymm. A tiny section of the original course of the river still remains in the grounds of Lymm Golf Club (with the concrete barges—see previous chapter).

In 1894, the Manchester Ship Canal opened with even more cuts and alterations made to the Mersey. The ship canal would use major sections of the Runcorn and Latchford Canal, the Mersey and the Irwell in its construction, the Runcorn and Latchford Canal being virtually obliterated with the exception of the Black Bear section which survived until the 1960s, running between Twenty Steps Lock on the ship canal to Manor Lock at Howley.

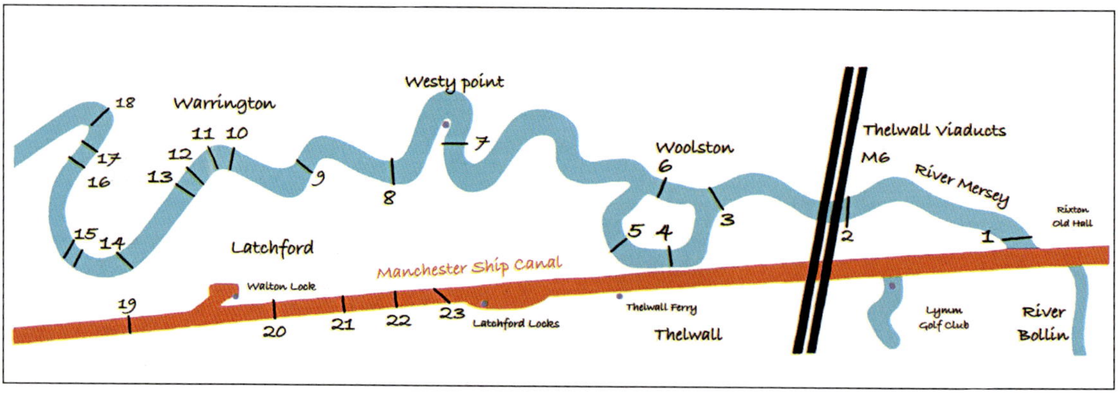


Map showing a section of the Manchester Ship Canal and River Mersey between Warrington and Rixton Old Hall.

This map represents the Woolston/Warrington section of the Mersey circa 2000 after the Thelwall Viaduct and New Woolston Weir had been completed. The map also shows the location of all the bridges that cross the Mersey and ship canal at this location.

## River Mersey 1–18

Manchester Ship Canal Rixton Railway Bridge
Thelwall Bailey Bridge
Woolston Guard Bridge
Woolston Footbridge
Woolston Aqueduct (water pipe)
Woolston New Weir
Paddington Meadows Aqueduct (water pipe)
Kingsway Bridge (Warrington)
Howley Suspension Bridge
Warrington New Bridge
Warrington Bridge
Arpley Railway Bridge
Arpley Blue Bridge
Centre Park Link Bridge (Slutchers Lane)
Twelve Arches Railway Bridge and Lower Walton Railway Viaduct
Loverose Way Bailey Bridge
Warrington Transporter Bridge
Bank Quay Aqueduct (Pipe)

## Manchester Ship Canal 19–23

Chester Road Swing Bridge
London Road Swing Bridge (alongside the disused Twenty Steps Lock)
Cantilever Bridge
Knutsford Road Swing Bridge
Latchford Railway Viaduct

Within the loop, over the Woolston Footbridge, was once the location of Wilgreave Farm, where the 'longest-living horse in the world'—Billy ('Old Billy')—was foaled and once ploughed the fields before being purchased by the Mersey and Irwell Navigation Company. Billy was born in 1760 and died in 1822; his skull is on display in Manchester Museum. The farm survived into the 1950s but closed when the land was used to dump dredgings from the ship canal. In fact, this tiny stretch of the Mersey has endured more cuts, deviations and alterations than any other location along the river (see map).

Before the Manchester Ship Canal was built, the loop in the Mersey once extended right to the most northerly point of Thelwall Village, next to where the current Thelwall Ferry is today. Thelwall itself is a charming village, with sandstone church, historic pub and post office. The sixteenth-century Pickering Arms bears an inscription that reads: 'In the year 920 King Edward the Elder founded a city here and called it Thelwall'.

*Above*: Woolston Footbridge.

*Below left*: Old Billy—his skull is on display in Manchester Museum.

*Below right*: Woolston Footbridge.

Back across the Guard Bridge and on to the Woolston 'new' weir, constructed in yet another cut in the Mersey. Preparatory work began in 1992 with the majority of construction taking place during 1993. The new 650-yard (600-metre) channel was dug at the top of the loop of bed 3 through the original route of the Mersey, diverted in the 1730s. At nearly 260 feet (80 metres) wide, the weir is a considerable structure and pioneering in its ability to effectively control and monitor water levels across the Mersey and Manchester Ship Canal for flood and navigation purposes. Three Archimedes screw turbines were added to the weir on the north bank to generate hydro-electricity and can generate enough electricity for nearly 500 homes.

The earliest weir at Woolston was built in the 1750s to provide a head of water to power a gunpowder mill that was built in 1758 but destroyed in 1855 following an explosion. The weir was located slightly south of where the Guard Bridge stands today, to the left of the powder mill.

The next weir was built in 1890 as part of the Manchester Ship Canal infrastructure by Sir Edward Leader Williams (engineer and designer of the ship canal, Warburton Toll Bridge, Barton Swing Aqueduct, etc.) to control water levels in the Mersey and the ship canal. The weir was initially manually operated and required full-time staff to lower or raise gates according to requirements.

The river continues its meander, leaving Woolston Eyes and taking another sharp turn around Paddington Meadows, alongside the 9-feet (2.7-metre) New Cut Canal (no longer in water), which once formed part of the Mersey and Irwell Navigation, cutting out the loops and earlier cuts in the Mersey around Woolston. The river passes beneath a further water pipe aqueduct and round Westy Point, a popular stretch of river for canoeing.

Thelwall Ferry.

*Above*: Woolston Weir with Archimedes screw turbines.

*Left*: Woolston Eyes.

*Right*: The Archimedes screw at Howley Weir.

*Below*: Woolston Weir.

*Above*: Westy Point.

*Below left*: Canoeing and rowing near Westy Point.

*Below right*: The Paddington Meadows Aqueduct, originally used to top up water in the Woolston New Cut Canal.

# 9

# Warrington

The river turns sharply west and passes beneath the Kingsway Bridge. Built in 1934, this double-arched bridge carries the A50 over the Mersey and is notable for its ten striking lanterns that adorn the bridge on each side, similar in design to coastguard buoys.

The Warrington Rowing Club is just beyond the bridge on the north bank of the Mersey, and on the south bank, almost directly opposite the rowing club, are the remains of Manor Lock that once joined the Black Bear Canal (also known as the Old Quay Canal) to the river and was used to transport animal hides to the tanneries at Howley.

Kingsway Bridge (Warrington).

The Black Bear Canal no longer exists but you can walk along its old course, through Latchford and under the Black Bear Bridge which carries Knutsford Road (A5061) over what was once the canal.

From the Kingsway Bridge, pedestrians can walk on either side of the Mersey and cross at the next footbridge, the Grade-II-listed Howley Bridge. David Rowell & Co constructed the bridge in 1912, primarily of wrought iron with steel cables and timber decking. The company built suspension bridges all over the world, including New Zealand, the Falkland Islands, Chile and the famous Queen's Park Bridge built in 1923 over the River Dee in Chester.

You can reach the Howley Weir on either side of the Mersey but it's worth staying on the north side of the river to check out the old Howley Lock. The river is tidal up to this point and a lock was required to allow shipping to pass upstream, although once the ship canal was built and traffic along the Mersey and Irwell Navigation dried up, the lock was no longer used and fell into disrepair. In 2021, H2O Power Ltd fitted a single Archimedes screw turbine that generates hydro-electricity in the disused lock. The weir was built in 1724 and changed the course of the Mersey, which once ran through Victoria Park in a huge loop as far as Wash Lane. An ancient ford once crossed the Mersey in the area, which is where Latchford gets its name ('Laeccford', which is Anglo-Saxon for a ford in a bog or stream). Boatmen named this stretch the 'hell hole' because boats would often get stuck in the shallow waters.

*Left and below*: Howley Bridge.

Warrington New Bridge (left) and Warrington Bridge (right) with the cenotaph in the centre.

About half a mile downstream and you will reach a sharp bend in the river—and a bridge lover's paradise. First up is the Warrington New Bridge, a road bridge built in 1985 to help ease congestion. Past the cenotaph on the south bank is the grand Warrington Bridge completed in 1915. The single-arch bridge was designed and built by Alfred Thorne and Sons and is the sixth bridge at this location, with the earliest on record dating back to 1285. The Battle of Warrington Bridge was fought at this site on 13 August 1651 between Charles II's Scottish Royalist forces and Major General John Lambert's Parliamentary army. The current Warrington Bridge replaced the Victoria

*Below left*: The capstone from the Victoria Bridge by the side of Mersey where the bridge once crossed the river.

*Below right*: The Warrington Bridge, completed in 1915.

*Above*: Arpley Railway Bridge.

*Left*: Arpley Blue Bridge.

Bridge built in 1837, the capstone of which sits by the footpath, close to the bridge, on the north bank of the river.

The Arpley Railway Bridge is up next, built in 1854 for the Warrington to Stockport Railway, engineered by John Lister from designs by Edwin Clark. No longer carrying passenger trains and rarely used at all, the bridge was a significant engineering achievement and unique in its wrought-iron box girder design. The line would once have carried trains over the Manchester Ship Canal using the now derelict Latchford Railway Viaduct.

At the next bridge, the aptly named Arpley Blue Bridge constructed by the Birse Group, the Mersey's original course took the river along Wilderspool Causeway in a curve through what is today Wilderspool Park. The river was diverted to a new straighter route, allowing larger shipping better access to Warrington via Walton Lock and the Manchester Ship Canal. Walton Lock was predominately used by Greenall Whitley & Co who operated their second brewery in Wilderspool from 1787.

The Arpley Blue Bridge carries Park Boulevard across the Mersey, built in 1990 to connect Chester Road to the Centre Park development. Cross the river at this point and continue along Chester Road towards the new Centre Park Link Bridge also known as Slutchers Lane Bridge, which opened to traffic in 2021.

Centre Park Link Bridge.

*Above*: The Twelve Arch Railway Bridge with the Lower Walton Bridge behind.

*Below*: The Lower Walton Railway Viaduct with the Twelve Arch bridge seen in front.

Continue along Chester Road (A5060) for 0.7 miles (1.12 kilometres) and take the small footpath on the right—this is another section of the Trans Pennine Trail which follows the old course of the Runcorn to Latchford Canal and leads to the Twelve Arch Railway Bridge. This magnificent stone-built bridge was completed in 1837 to carry the Grand Junction Railway from Warrington to Birmingham. The neighbouring railway bridge was constructed in 1890 as the line was required to reach a much greater height in order to cross the still-under-construction Manchester Ship Canal. Acton Grange Viaduct was constructed to take the line over the ship canal—once the new line was complete, the original line was gradually raised to reach the height required to reach the viaduct, therefore avoiding a long period when the line would have been out of use. Both bridges are still in use and would have once crossed the Runcorn to Latchford Canal, as well as the River Mersey.

The path continues beneath the two railway bridges and takes a left across Arpley Meadows as the road ahead leads to private land and various industrial and chemical plants. However, no journey to Warrington is complete without viewing the Warrington Transporter Bridge and there is a little-known footpath that will take you to it.

If you know Warrington at all, you'll be familiar with the Pink Eye Building. Once a grain store for the Fairclough Flour Mill, the building was painted by a local artist Anthony Turk at the request of the owner—when the council refused permission to paint a sealife mural on it, the artist painted it pink instead. The building has become a local landmark and can easily be found off the Liverpool Road in Bank Quay. Across the railway level crossing on Quay Ford, a path leads through the chemical works, past shipping containers and an industrial aqueduct, along a walkway adjacent to Factory Lane and eventually to the Warrington Transporter Bridge.

Lower Walton Railway Viaduct.

*Above left*: Lower Walton Railway Viaduct.

*Above right*: Looking through Lower Walton Viaduct towards Arpley Meadow.

*Left*: Lower Walton Railway Viaduct.

*Opposite above*: Warrington Transporter Bridge with the Bailey Bridge visible in the distance.

*Opposite below*: The gondola was equipped to transport railway trucks across the River Mersey, the first of its kind.

The Warrington Transport Bridge was constructed in 1916 by William Arrol & Co from designs by William Henry Hunter. It was the second transporter bridge at this location; the first, which no longer exists, was built in 1905 slightly further north of the current bridge. The bridge was designed to take railway wagons across the Mersey, where freight wagons could be wheeled directly onto the gondola without the need to unload. The gondola was converted to road vehicles

The Warrington Transporter Bridge was built for Joseph Crossfield's soap works.

in 1940 but sadly fell into disuse in the late 1960s. The bridge is a Grade II listed building and is also on the Heritage at Risk Register. Transporter bridges were once quite a popular design and a relatively cheap way of transporting people, goods or vehicles across a river while allowing tall ships to pass beneath with relative ease. Along with the Woolston Footbridge, the Warrington Transporter Bridge was nominated for the Institution of Engineering North West Heritage Award in 2016.

A further Bailey bridge crosses the river south of the transporter bridge and carried HGVs across the Mersey, making the transporter bridge redundant and contributing to its ultimate demise.

Joseph Crosfield, known locally as 'Joe Soaps', established his soap and chemical works on the banks of the River Mersey in 1814 at Bank Quay, although the area was largely fields at the time. The Joseph Crosfield & Sons Ltd offices are still visible on Liverpool Road, close to a war memorial erected by the Crosfield family commemorating employees killed or lost during the two world wars. Close to Joseph Crosfield's business, William Lever set up his first soap-manufacturing business at Bank Quay in 1885, running the company with his brother under the name Lever Brothers.

Warrington Bank Quay Aqueduct.

*Above*: Warrington Transporter Bridge.

*Below*: Warrington Transporter Bridge.

Forrest Way Bridge.

As the Mersey starts to leave behind Warrington's historic manufacturing past, foliage replaces the factories that dominated the riverbanks round Bank Quay. Warrington will present one last bridge to cross in the form of the Forrest Way Bridge—built in 1987, although made from reinforced concrete, its steep arch and contours are surprisingly pleasing to the eye.

For the next five miles or so, the river transforms, broadening and starting to become truly affected by the tides, as sandbanks become evident around each curve. The depth of the river increases and decreases as the Mersey slowly starts to breathe the Irish Sea.

Moore Nature Reserve occupies the south bank of the Mersey, and the north bank flows alongside a section of the Sankey Canal, still in water, and the St Helens and Runcorn Gap Railway.

Fiddler's Ferry lent its name to the nearby power station but actually relates to the Norton Priory monks who operated a ferry in the twelfth century. The Sankey Canal was extended to Spike Island in 1762 and contains a lock to the Mersey, still in use today. The railway followed in 1853 with a station named Fidler's Ferry and Penketh, which was renamed Fidler's Ferry and Penketh Station in 1881. The station closed in 1963 and today the line only carries freight.

*Above*: View towards Bank Quay with the transporter bridge centre right.

*Below*: The Ferry Tavern at Fiddler's Ferry.

The Ferry Tavern (originally the Ferry Inn) opened in 1762, the same year as the canal extension. Popular with walkers and cyclists, it's a fantastic place to sit outside in the warmer months with a pint and watch the river ebb and flow. Seals and porpoises have been photographed along this stretch of the river and a fish survey carried out in 2023 reported a record forty-five species of fish now thriving in the Mersey's waters.

The neighbouring Fidler's Ferry Sailing Club offers a wide variety of water sports and hosts dinghy races and canoeing trips all supervised from the observation tower, which was originally the wheelhouse of a seagoing vessel.

Moore Nature Reserves sits alongside the south bank of the river, nestled snugly between the Mersey and the ship canal. Once farmland and a sand quarry, the area now has five lakes, woodland and several bird-watching hides. A bridge close to the entrance of the reserve once carried the Manchester Ship Canal Mineral Railway line and is a near-identical design to the bridge crossing the River Mersey at Rixton.

The footpath on the north bank continues along the Sankey Canal but bypasses an interesting bend in the Mersey where the river brushes alongside the Manchester Ship Canal. The river curves back and rejoins the path and at this point there's an interesting 'wind-powered' sculpture called Future Flowers, which signifies the transformation from industry to nature, before the river broadens once more and reveals the Mersey Gateway Bridge.

The tug *Bonchurch* moored at Fiddler's Ferry Marina.

The Sankey Canal with the Fiddler's Ferry Power Station in the background.

10

# Runcorn

Construction on the Mersey Gateway Bridge began in May 2014 and was finally opened to traffic on 14 October 2017. Not the most recent bridge to be built across the Mersey, but it is the largest, with an overall span, including approach viaducts, of 1.4 miles (2.3 kilometres), and it carries three lanes of traffic in each direction. The bridge has three towers, each with supporting cables in a distinctive harp-shaped design. Knight Architects designed the bridge and are responsible for many innovative bridges across the world

Mersey Gateway Bridge with views across the River Mersey to Spike Island.

*Above left*: Looking across the Mersey to Moore Nature Reserve.

*Above right*: Mersey Gateway Bridge.

*Below*: Mersey Gateway Bridge with views across the River Mersey to Spike Island.

including the Ordsall Chord in Manchester. When seen for the first time, the bridge will simply take your breath away, particularly if viewed from either Spike Island or Wigg Island, both of which are community parks and nature reserves reclaimed from industry that sit either side of the bridge.

Wigg Island, on the south bank of the river, was originally a saltmarsh before the Manchester Ship Canal was built, effectively creating a strip of land between the ship canal and river. The island is named after Charles Wigg, who built his factory 'Wigg Works' in the 1860s to manufacture alkali. The Manchester Ship Canal replaced the Runcorn to Latchford Canal but brought with it the Manchester Ship Canal Railway that ran along the north bank of the ship canal and could transport goods to and from Warrington. The island contains a small section of the Runcorn to Latchford Canal that is still in water.

Charles Wigg sold his alkali plant in 1890 to the United Alkali Company, which was subsequently acquired by ICI in 1926 and the site was greatly developed and expanded to manufacture other chemicals. The factory was a major manufacturer of mustard gas during the Second World War but all chemical production ended in the 1960s and the area was left heavily polluted.

Today, the area is a nature reserve and community park which opened in 2002, accessible via the Old Quay Swing Bridge off Mason Street in Runcorn. I only became familiar with the area early in 2022 but, along with Spike Island, across the river, it is an area I have greatly enjoyed visiting and find myself repeatedly being drawn back to.

The Mersey Gateway Bridge.

The Old Quay Swing Bridge providing access to Wigg Island.

From Wigg Island, staying on the south side of the river, the Manchester Ship Canal now runs alongside the Mersey, separated either by a tiny strip of land or a purpose-built stone barrier. The disused Old Quay Lock sits halfway between Wigg Island and the Jubilee Bridge and once connected the Runcorn to Latchford Canal to the River Mersey.

Along the Runcorn Promenade, close to the Silver Jubilee Bridge, there is a short pier which was the location of another iconic bridge, the Widnes–Runcorn Transporter Bridge, built in 1909 but closed in 1961 and later demolished. The gondola or cradle was much larger than the one on the Warrington Transporter Bridge and was designed to carry up to four horse-drawn carts and 300 people, although the crossing was very time-consuming and the bridge was often in need of repair and unable to meet the needs of the region. A sad loss that could have been a great tourist attraction today, the spike from the roof of the gondola is displayed in the car park of the Catalyst Science Museum in Widnes.

The Runcorn–Widnes Bridge was built alongside the transporter bridge, and opened by Princess Alexandra on 21 July 1961. Later widened in 1975–77 and renamed the Silver Jubilee Bridge in honour of Queen Elizabeth II's jubilee in 1977. The middle span is 1,082 feet (330 metres) wide and its height over the river bed is 285 feet (87 metres). This Grade II listed building was the largest steel-arched bridge in Europe at the time of its construction and still remains a major local landmark.

*Above left*: The short pier was the location of the Widnes–Runcorn Transporter Bridge.

*Above right*: View from Wigg Island towards Runcorn's Silver Jubilee Bridge—note the difference in colour between the Manchester Ship Canal (left) and the tidal River Mersey (right).

*Below*: The Silver Jubilee Bridge with the Runcorn Railway Bridge behind it, barely distinguishable at this angle.

The River Mersey famously divided Lancashire (north of the river) from Cheshire (south) and centuries before, the kingdoms of Northumbria (north) from Mercia (south). Alfred the Great's eldest daughter Ethelfleda (Æthelflæd), Lady of the Mercians, ruled Mercia from 911 until her death in 918. She was responsible for building and strengthening many burhs (fortified settlements) across western Mercia and up to the northern border of the kingdom at Runcorn. All Saints' Church is thought to have been built at the location of an earlier church erected by Ethelfleda as part of the settlement constructed by her to defend against the Danes.

Ethelfleda lends her name to the Runcorn Railway Bridge (the bridge with three names, as it is also known as Ethelfleda Bridge or Britannia Bridge), which runs alongside the Silver Jubilee Bridge. This was the first bridge to cross the Mersey at this location. A ferry had crossed the Mersey in Runcorn from as early as 1178. A blue plaque commemorating the ferry crossing is mounted on the Boat House, previously called the Ferry Boat, on Mersey Road and is close to the location of the ferry. Immortalised in rhyme by Marriott Edgar, 'The Runcorn Ferry' was spoken over music and made famous by Stanley Holloway.

The Runcorn Railway Bridge was built in 1868 from designs by William Baker and consists of six lattice girders over three wrought-iron spans, supported by two

The Runcorn Pier, once the location of the Widnes–Runcorn Transporter Bridge.

*Above*: The Silver Jubilee Bridge by night.

*Below*: Old Quay Lock was originally part of the Runcorn and Latchford Canal (also known as the Old Quay Canal).

sandstone abutments built on foundations 45 feet (14 metres) deep. In Runcorn, the bridge is approached via a thirty-three-arch viaduct, but in Widnes the bridge's approach boasts an even more impressive forty-nine arches. The bridge also incorporates a public walkway, sadly closed in 1965 but still in use for maintenance. The official name of the bridge is the Runcorn Railway Bridge but as mentioned, it is also known locally as the Britannia Bridge; this is due to a series of mounted shields on the bridge which bear the crest of the railway company and incorporates the images of Britannia and Ethelfleda, owing to the belief that the southern foundations of the bridge were constructed at the site of her burh.

A lock is still visible in the barrier between the Manchester Ship Canal and the River Mersey, called the Old Quay Lock. Located opposite The Decks Apartments, near Mersey Road in Runcorn, it was originally part of the Runcorn to Latchford Canal, also known locally as the Old Quay Canal. A few small sections of the Runcorn to Latchford Canal still exist and can be viewed in water at Wigg Island and Moore Nature Reserve, but this section alongside the Old Quay Lock was repurposed and became the Manchester Ship Canal. The Decks Apartments were built on the quayside of what was once the location of the Old Quay Shipyard. The shipyards and workshops survived the development of the ship canal and although no ships were built at the site after 1890, the yard became the centre for repairing the tugs and lock gates of the Manchester Ship Canal Company, which continued until 2003.

New apartments (far left) built at the location of the Old Quay Shipyard.

*Above*: Slipway No. 2 is the last remnant of the Old Quay Shipyard.

*Below*: The Runcorn Railway Bridge, also known as Ethelfleda Bridge or Britannia Bridge.

*Above left*: The Runcorn Railway Bridge with The Decks Apartments and Old Quay Lock visible between the arches.

*Above right*: The Runcorn Railway Bridge at low tide.

I personally like the Queen Ethelfleda notion, but whichever you prefer, it cannot be denied the bridge is stunning and a huge engineering achievement. From a distance, the two bridges do appear to merge and the magnificence of the railway bridge is lost against the spectacle of the Silver Jubilee bridge—but get up close and you will appreciate its grandeur.

Beyond the bridge, the Mersey curves south and starts to broaden once again. The Manchester Ship Canal passes Weston Point Docks and is joined by the River Weaver.

# 11
# Widnes

Returning to the north side of the Mersey, we left the river at the Merseyside Gateway Bridge. Like Wigg Island on the south bank, Spike Island has an industrial past and shares its remarkable transformation from a contaminated, decaying wilderness to a pleasant, clean community park.

Spike Island is bordered by the Mersey on the south side with the Sankey Canal curving round the north and western corners. Spike Island was the world's first rail-to-ship dock and became the centre of chemical manufacturing in Britain during the Industrial Revolution. No trace of railway lines exist today but the island once had tracks bordering the canal, around the dock and factories. The railway officially

Spike Island was the world's first rail-to-ship dock.

opened in February 1833, shortly before the canal in July, and finally the dock in August.

John Hutchinson built a number of chemical factories on Spike Island and the nearby Tower Building, his administration offices, which still stands today and is the home of the Catalyst Science Museum.

Spike Island was the location for an outdoor concert by the Stone Roses on 27 May 1990, attended by more than 28,000 fans. The concert inspired a 2012 film called *Spike Island*, which follows a group of dedicated but ticketless fans intent on attending the gig. Original footage of the 1990 concert was discovered in 2024 and is due to be released as part of a documentary about the event.

Varying water levels and shifting sandbanks made it difficult for larger ships to navigate the Mersey estuary and goods would be transported to and from the Port of Liverpool on barges called 'Mersey Flats'; these were also small enough to navigate the canals and waterways much further upstream, as well as the docks at Widnes, Runcorn and the Mersey and Irwell Navigation.

The Mersey Flat was an iconic flat-bottomed barge, traditionally made with a timber frame, doubled-ended with a single mast. The barges were used from the early eighteenth century on the Mersey, Irwell and Weaver rivers as well as the surrounding

Widnes with the Jubilee Bridge centre right.

*Above*: Spike Island Dock.

*Below*: Mersey Flat wrecks litter the shoreline around Spike Island.

*Above left*: *Mossdale* at the National Waterways Museum.

*Above right*: The abandoned skeletal remains of boats at Spike Island.

*Below*: The former powerhouse of the Widnes–Runcorn Transporter Bridge.

*Above left*: The tunnel once led to the loading pier.

*Above right*: St Mary's Church in West Bank.

canals. The skeletal remains of a couple of Mersey Flats can be found around the eastern bank of Spike Island at low tide.

A second dock was built at Widnes in 1864 called the West Bank Dock and initially comprised two areas of land, Widnes Marsh and Hurst Marsh. A third section was added in 1865 along the shore on the River Mersey. The massive dock complex included an extensive rail network, factories, warehouses and canal link.

The dock continued to expand and by 1948 covered a colossal 280 acres. Today, the Victoria Promenade runs along the bank of the river with views of Runcorn across the Mersey and St Mary's Church inland. The church features an unusual outdoor pulpit, built into the southeastern boundary wall.

Towards the end of the promenade is the powerhouse for the Widnes–Runcorn Transporter Bridge, which outlasted the bridge itself and is a Grade II listed building today. Just beyond the powerhouse, on the banks of the Mersey, are the remains of the ferryman's hut.

The Catalyst Science Museum on Mersey Road has a permanent Widnes-Runcorn Transporter Bridge exhibition and even has the decorative spike from the roof of the bridge's gondola on display in the car park.

12

# Wirral

Across the Mersey, the river squeezes through Runcorn Gap and turns southward to pass Runcorn Docks and Weston Point, where the River Weaver joins the Manchester Ship Canal. Heading westerly once more, the river broadens and still shadowed by the Manchester Ship Canal, passes Frodsham, Ince, Stanlow Oil Refinery and Ellesmere Port.

Staying alongside the Mersey or ship canal at this point is impossible due to the sheer volume of industrial sites, docks and ports. As we journey towards the Irish Sea, we can explore a few locations that offer a view of the Mersey or are of historical interest.

*Opposite above*: View from Frodsham Marsh with Stanlow in the distance and the wind farms on the far left.

*Opposite below*: Aerial view of Stanlow.

*Below*: Stanlow Oil Refinery.

Originally a small bitumen plant built in 1924, Stanlow Oil Refinery now occupies a 1,900-acre site and produces over 15 per cent of the UK's petrol.

The mysterious Stanlow Island, a long strip of land that sits offshore alongside Stanlow and Ellesmere Port, was cut off from the mainland when the Manchester Ship Canal was created. The island, inhabited until the early 1990s, is home to the Stanlaw Abbey—Stanlaw being the original spelling until an error was made on a Victorian

*Above*: View of Stanlow and the wind farm from Eastham.

*Below*: Stanlow Island showing the River Gowy flowing beneath the Manchester Ship Canal and across the island to meet the River Mersey.

map. The Abbey of St Mary at Stanlaw was founded in 1178 and played host to King Edward I when he stayed there in August 1277.

The abbey was partially destroyed by storms and floods in the late thirteenth century and the majority of the monks relocated to Whalley near Clitheroe in 1296, although a small group remained until the Protestant Reformation in the late sixteenth century. Little remains of the abbey today, with the exception of some of the foundation stones. The houses were demolished shortly after the last inhabitants left the island in the 1990s but the police station is still visible, once home to the Manchester Ship Canal Police Force.

The River Gowy flows through a syphon beneath the Manchester Ship Canal and across the island for just under half a mile (0.8 kilometres) to meet the River Mersey. Most of the island is leased to the Indian company Essar Energy, who own the Stanlow Oil Refinery and the docks located on the island.

Downstream from the oil refinery is the National Waterways Museum, which occupies the site of the historic Netherpool Port, built in the late eighteenth century and designed by Thomas Telford. The port enabled cargo to be transferred between canal, river and, subsequently, the docks at Liverpool. Since the 1970s, the seven-acre site has been the home of the North West Museum of Inland Navigation, later renamed the National Waterways Museum, and has been managed by the National Waterways Trust since the 1990s. The site has a vast collection of historic boats and river craft, including a magnificent Mersey Flat—the Mossdale, built in 1876 by the Shropshire Union Railway and Canal Company, originally named Ruby. Abel and Son in Runcorn purchased, refitted and renamed the boat in the 1930s.

Stanlow Island and Stanlow Port.

*Above*: Manchester Ship Canal cuts through the landscape to create Stanlow Island.

*Below*: National Waterways Museum.

The Shropshire Union Canal enters the Manchester Ship Canal via locks at the National Waterways Museum, adjacent to which is located the Ellesmere Lighthouse, also known as the Whitby Lighthouse and the last remaining of three that used to illuminate the southern coastline of the Mersey, the others being at Ince and Weston. Built in 1880 at the site of a previous lighthouse, the Ellesmere Lighthouse was no longer used when the Manchester Ship Canal was completed in 1894.

The modern Ellesmere Port Dock is situated just beyond the lighthouse and was brought back into use in 2007 following an increase in bulk cargo transportation.

The Mersey is at its widest point here, measuring roughly 3 miles (4.8 kilometres) between Ellesmere Port and the Speke Hall Estate. The Mersey has the second highest tidal range in the UK: the river can rise by as much as 32 feet (10 metres). Many sandbanks are exposed at low tide including the treacherous 'Dungeon Banks' between Hale and Stanlow.

At Eastham, the Manchester Ship Canal reaches its conclusion, completing its 36-mile journey from Salford Quays. In the nineteenth century, Eastham, once the site of a ferry service from Liverpool, was a popular holiday destination with ornamental gardens, boating lake, ballroom, bandstand, tea rooms and a small zoo with bears, lions and monkeys. An iron pier extended from the stone jetty, which, by 1816, enabled paddle steamers to dock, replacing the earlier sailing boats that were used to ferry passengers across the Mersey.

*Below left*: Ellesmere Port.

*Below right*: *Stolt Seagull* passing Ellesmere Port.

*Above*: *Patara* leaving Eastham Lock.

*Below left*: The abandoned Bear Pit, once a feature in Eastham Pleasure Park Gardens.

*Below right*: Remains of the Eastham Iron Pier visible at low tide.

*Above*: The location of Job's Ferry.

*Below*: View from Job's Ferry.

At low tide, the remains of the iron pier can be seen in the sand from the stone jetty, which is still intact and offers great views of the Mersey, Liverpool and shipping entering the locks at Eastham, the fifth and final set of locks along the ship canal. Slightly further north of the jetty lie the remains of the original Eastham ferry, known as 'Job's Ferry', which was operated by monks from St Werburgh's Abbey in Chester from 1357; the ferry was predominately used to transport goods and farm produce from Chester and North Wales across the Mersey to Liverpool. Steps cut into the sandstone leading down to the Mersey are clearly visible and form the landing site of the ancient ferry.

View from Eastham towards Liverpool.

Port Sunlight.

Heading towards Birkenhead, we pass Port Sunlight, the historic and picturesque village created by William Hesketh Lever for workers at the Lever Brothers Soap Works, an expansion of his successful soap manufacturing business in Warrington. In 1888, Lever purchased 85 acres of marshland on the Wirral Peninsula and tasked over thirty architects to design and build different elements of the village, which would incorporate a new manufacturing plant, workers' cottages, parks and gardens, school, cottage hospital, church, theatre, concert hall, public house, boating pond and art gallery. By 1914, the village housed 3,500 employees and their families in 800 comfortable homes, each with hot and cold running water, in beautiful parkland. A very radical approach at that time, Lever's aim was to invest in his employees as he considered happier and healthier workers would make a more productive workforce.

The village is a sumptuous delight of architecture, monuments and gardens, with over 900 Grade II listed buildings—there is something to marvel at around every corner. Lever was a visionary and his intentions were revolutionary—although he failed to take the same approach with his African workforce.

In 1911, Lever set up a company with the Belgian government, in the occupied Belgian Congo, to produce palm oil, a raw ingredient for soap manufacturing. Lever's initial intention was to build a second Port Sunlight, called Leverville, for his Congolese workers, but insufficient numbers of workers were recruited voluntarily, so the barbaric and often brutal Belgian practice of forced labour was employed.

*Above and below*: Beautifully designed workers' houses at Port Sunlight.

Ship Building at Birkenhead.

Birkenhead is perhaps best known for shipbuilding, which is still in evidence today along the west bank of the Mersey. Starting in 1824, William Laird founded an iron works to manufacture boilers. By 1828, in partnership with his son John, he set up a new business and they built their first iron barge, going on to make pioneering advances in shipbuilding, design and propulsion. In 1903, Laird and Sons merged with the Sheffield-based Johnson Cammell & Co to form Cammell Laird, which would be renowned for shipbuilding and ship repair for more than 120 years.

The remains of Birkenhead Priory are adjacent to the Cammell Laird shipyards and it is the oldest building on the Wirral Peninsula, founded in 1150 by the 3rd Baron of Dunham Massey. The priory's chapter house has been extensively restored and includes a small chapel dedicated to the sailors lost aboard HMS *Conway*, built at Cammell Laird Shipyard but sank in the Mersey during trials. The site also encompasses St Marys Church, Birkenhead's first parish church, which opened in 1821. The tower of the church is fully intact and visitors should make the effort to climb to the top for stunning views across the River Mersey and Liverpool beyond. The church lost its graveyard in the 1960s when the neighbouring Cammell Laird expanded its dry dock.

Richard I is reported to have visited the abbey on several occasions and Richard II granted ferry rights to the monks of the abbey who provided shelter for travellers, as well as the first 'approved' ferry across the Mersey.

*Above left*: Shipyards.

*Above right*: Cammel Laird Shipyards.

*Below*: View from the top of St Mary's Church in the grounds of Birkenhead Priory.

*Above left*: Birkenhead Priory with St Mary's Church on the right.

*Above right*: The chapel at Birkenhead Priory.

*Below*: Birkenhead Town Hall.

Birkenhead can boast many 'firsts': the first street tramway in Britain, the world's first railway tunnel under a river estuary, and the first publicly funded park in Britain—Birkenhead Park, designed by Joseph Paxton, which inspired Frederick Law Olmsted's designs for Central Park in New York.

William Laird commissioned Scottish architect James Gillespie Graham to design a town square in Birkenhead. Construction on Hamilton Square began in 1825, with

Hamilton Square.

the area's first market square being added in 1835 and, eventually, Birkenhead Town Hall, completed in 1887. The square is a pleasant and peaceful corner of Birkenhead and home to the most Grade I listed buildings outside Trafalgar Square in London.

Hamilton Square is also the name of the railway station closest to the river, linking Birkenhead to Liverpool via a tunnel under the Mersey completed in 1886. The tunnel initially used steam locomotives to pull wooden carriages the 3.12 miles (5 kilometres) between Hamilton Square Station and James Street in Liverpool. Although an elaborate ventilation system had been installed, the tunnels were heavily polluted with fumes and the lines were subsequently electrified in 1903. One of the steam locomotives used in the Mersey Railway Tunnel is in storage at the Museum of Liverpool: the 'Cecil Raikes' was built in Manchester by Beyer Peacock and Co Ltd.

There are no bridges across the Mersey between Liverpool and the Wirral but there are three tunnels: the aforementioned railway tunnel and two later road tunnels.

The Queensway Tunnel was opened in 1934 by King George V, after nine years of construction, and at 2.01 miles (3.24 kilometres) in length it was the longest road tunnel in the world until the Vielha Tunnel was completed in Spain in 1948. The tunnel includes an upper deck to carry four lanes of road traffic, a lower deck

Kingsway Mersey Tunnel at Wallasey.

originally conceived to carry a tramway, but today used as an emergency refuge, and two ventilation shafts. Herbert James Rowse designed the tunnel entrances, the ventilation building exteriors and the tollbooths, which contain some stunning art deco detail, as well as other notable buildings in Liverpool including the George's Dock Building and Philharmonic Hall. The tunnel runs directly beneath the Woodside Ventilation Tower, which is a landmark on the riverbank in Birkenhead and lavishly decorated, winning Rowse the Bronze Medal from the Royal Institute of British Architects in 1937.

The Kingsway Tunnel comprises two identical 1.5 mile-long (2.4 kilometres) tunnels, each containing two lanes of traffic that run from Liverpool to Wallasey. The southern tunnel was completed first and opened by Queen Elizabeth II on 24 June 1971, the northern tunnel was completed in 1973 and opened to traffic on 13 February 1974. There are additional service tunnels that connect the two for maintenance purposes.

The Wallasey Pool was a tidal inlet that separated Birkenhead from Wallasey, which in itself is almost an island, being bordered by the River Mersey, the Irish Sea, Wallasey Pool and dense marshland. The name derived from Anglo-Saxon, meaning 'Island of the Welsh or Britons'. From 1842, docks were built within Wallasey Pool and the area became dominated by engineering, manufacturing and sugar-refining industries.

*Below left*: Queensway Mersey Tunnel in Liverpool.

*Below right*: The monument to the Mersey Tunnel—the final remaining light pillar designed by Herbert James Rowse, located close to the western entrance of the Queensway Tunnel.

*Right*: One of the original Queensway Mersey Tunnel tollbooths by the George's Dock Building in Liverpool.

*Below*: Mersey Ferry—an iconic image.

Ferries across the Mersey have a long history and enjoying the River Mersey by boat should be on everyone's bucket list. You can catch a ferry from the Seacombe Ferry Terminal to Liverpool, or the more adventurous might like a Liverpool Bay Cruise or even a sixty-seven-hour cruise along the Manchester Ship Canal. The Mersey does look quite murky in Liverpool and around the Wirral Peninsula but this is due to the river being tidal and the huge sandbanks that are exposed at low tide across the estuary from Runcorn up to Eastham. The river is also continually dredged to maintain the shipping lanes, which also increases the water's turbidity.

From Seacombe Ferry Terminal, you can once again enjoy the River Mersey by foot and walk or peddle the final 2.5-mile (4 kilometres) stretch of the riverbank to New Brighton. Split into three, Seacombe Promenade contains Wallasey Town Hall, Egremont Promenade has the remains of the old Egremont Pier, demolished in 1946 following a shipping accident, and finally, Magazine Promenade which leads directly to New Brighton, Fort Perch Rock and the end of our journey along the Wirral Peninsula.

The River Mersey with Seaforth cranes top left.

*Above*: View from the beach along Seacombe Promenade beach towards Seaforth.

*Below*: New Brighton Lighthouse and Seaforth Dock.

Fort Perch Rock sits at the northernmost tip of the Wirral Peninsula, where the Mersey finally meets the Irish Sea. Built between 1825 and 1829 to protect the developing Port of Liverpool, with room to house a hundred men, and armed with eighteen guns, the fort originally had a drawbridge and was cut off at high tide. Designed by Captain John Kitson, the fort was decommissioned by the War Office in 1956 and today houses a permanent museum, which features a replica of the Titanic radio room and an exhibition about the HMS *Thetis*, the ill-fated Cammel Laird-built submarine. The fort was to include a lighthouse but plans changed and the Perch Rock Lighthouse (renamed New Brighton Lighthouse) was built in 1827 behind the fort closer to the sea.

The fort and lighthouse are now famous landmarks of New Brighton and are permanent reminders of the past and the part they played in the area's history—a fitting end to our journey along the Wirral Peninsula, but there's more to see 'cross the Mersey in Liverpool.

Seaforth Docks by night.

*Right*: New Brighton Lighthouse and Fort Perch Rock centre right.

*Below*: Liverpool from Perch Rock.

13

# Liverpool

On the north bank of the Mersey, we continue our journey from Widnes at the West Bank Dockland Park, with views of the Runcorn Railway Bridge and across the river to Weston Point to the pleasant nature reserve and flower meadows of Pickering's Pasture.

As the river broadens and bends around Hale, strong tides, hidden undercurrents and shifting sands present treacherous conditions for shipping. In 1838 a lighthouse was built on the sandstone cliffs at Hale Head, replaced by a taller, sturdier lighthouse in 1906 that lit the skies and warned sailors of the perils that lay ahead. Decommissioned in 1958 as shipping transferred to the Manchester Ship Canal, the lighthouse remains a historic local landmark and is a Grade II listed building.

Hale Lighthouse.

*Right*: Statue of John Middleton, the Childe of Hale.

*Below*: Speke Hall.

*Above*: Hale Lighthouse from Frodsham Marsh. Can you spot the plane about to land at John Lennon Airport?

*Opposite*: A view of the Mersey and Stanlow beyond from Oglet.

Hale Village, a short walk north of the lighthouse, is a charming little hamlet with beautiful, thatched cottages, manor house and historic church with a tower dating back to the fourteenth century. The church graveyard is the resting place of John Middleton, known as the Childe of Hale, whose bronze statue dominates the village green. Reputed to be 9 feet 3 inches tall, John was once challenged to wrestle King John I's champion and won, breaking his arm for good measure.

A view across the Mersey from the Festival Gardens.

The river curves in an 's' shape around rural farmland and a slip of land called Oglet that sits between the Mersey and John Lennon Airport, where flint artefacts have been found possibly dating back to the Mesolithic or Bronze Age. Roman coins and a brooch were also found at the location, so the site may have been a settlement at some point in history.

Beyond the runway of the airport lies the Speke Hall Estate. The striking timber-framed Tudor moated house is set in beautiful gardens and woodlands, with the Great Hall dating back to 1530 when it was constructed under Sir William Norris. William was succeeded by his son Edward, who completed the gatehouse in 1598.

In the mid-seventeenth and early eighteenth centuries the Norris family obtained much of their wealth through their connections with the slave trade, and as MPs were responsible for greatly expanding the trade in enslaved Africans.

Speke was sold to Richard Watt in 1795, who invested in three slaving voyages in the late eighteenth century, owned and managed his own plantation using enslaved workers, and was responsible for transporting 539 African people to Jamaica.

I would recommend a visit to Speke Hall; the grounds and hall are magnificent but be aware of the history surrounding some of its occupants and how they acquired their immense wealth.

Alongside an industrial estate sits the Port of Garston which operates as a shipping and container port, beyond which is the Otterspool Nature Reserve and start of the Otterspool Promenade, a footpath and cycle lane that continues along the north bank of the Mersey and into the heart of Liverpool.

Otterspool leads into the Liverpool Festival Gardens, a pleasant stretch that was developed in 1984 for the International Garden Festival to boost tourism, although following the initial festival, some of the original site has been used for a new housing development. It's a pleasant area for a walk along the river, although some of the original Festival Garden features are in need of refurbishment.

Beyond the festival gardens, the promenade reaches an area that was once the Herculaneum Docks, built in 1866, and the start of the impressive 7.5 mile-long (12 kilometres) network of Liverpool Docks. Herculaneum was built to handle petroleum, which was stored in the sandstone cliffs overlooking the docks and used during the Second World War for the North Atlantic Convoys. The docks closed in 1972 and were filled in during the 1980s—with the exception of dock 4, which was maintained as a feature for a housing development.

The docks were also the location of the Liverpool Overhead Railway Southern Extension, completed in 1896, and served the dock with its own station. The names of

Liverpool Festival Gardens.

*Left*: Otterspool Promenade.

*Below*: Liverpool Promenade alongside Riverside Drive and the location of the Herculaneum Docks.

the chairman, engineers and contractors can still be seen engraved into the brickwork of the tunnel.

Harrington and Toxteth Docks, neighbouring docks to Herculaneum, were both closed in 1972 and filled in during the 1980s.

Brunswick Dock has a lock gate connecting it to the River Mersey and was originally used as a timber dock. Today, along with the adjacent Coburg Dock, it is used as a marina with 340 berths for private yachts, canal barges and other sailing vessels. Brunswick Dock connects to all docks on the Liverpool waterfront and the Leeds-to-Liverpool Canal via Stanley Dock.

Liverpool was originally little more than farmland with no major settlements until it was granted a royal charter in 1207 by King John making it a borough, and was then used as a site to transport men and materials across the Irish Sea for his campaigns in Ireland. A castle was built shortly afterward in 1230 with a small network of seven streets running from the west side of the castle and down to the River Mersey. To the east of the castle was a natural, tidal inlet called the 'Lyver Pool', which expanded at high tide but drained at low tide, making the perfect location for Thomas Steers to build the world's first commercial wet dock, completed in 1716.

A picturesque housing development constructed around the final remaining dock (no. 4) from the Herculaneum Dock complex.

*Above*: The petroleum sheds built into the sandstone cliffs behind the Herculaneum Docks.

*Below*: Liverpool Overhead Railway Southern Extension.

Liverpool historic docks.

Liverpool was already developing as a port but shipping was restricted by the tides of the Mersey and a more robust solution was required if the port was to expand and prosper. The Thomas Steers Dock (now better known today as the Old Dock) took five years to build, at what was then a phenomenal cost of £12,000, almost bankrupting Liverpool—but the gamble paid off and trade flourished.

As Liverpool grew and the waterfront expanded, larger docks were built along the Mersey, reclaiming land back from the river, leaving the Old Dock redundant and eventually built over. During an archaeological excavation that began in 2001, a corner of the historic Old Dock was excavated and now forms part of a guided tour. Now hidden beneath the Liverpool One shopping centre, the dock still had a few secrets left to reveal and a bricked-up culvert was discovered in the sandstone at the lowest part of the dock, pre-dating Thomas Steers's construction and thought to be a secret entrance to the castle. Stone from the ruins of the castle were also thought to have been used in the construction of the dock itself.

William Hutchinson, an extraordinary privateer captain, inventor, author and sailor once held the position of Dock Master at Liverpool and its expanding dock network. Already an experienced seaman, William was offered the post in 1759 and accepted. His exploits took him around the world, taking part in many sea battles and government-sanctioned raids on French and Spanish shipping. He survived a shipwreck and was even once captured and imprisoned by French authorities before joining the Royal Navy in 1743. William's adventures continued on land and he designed what may have been the first parabolic mirror for the Leasowe Lighthouse, helped establish the Liverpool Marine Society for impoverished sailors, invented a new ship rudder and cofounded what is believed to be the world's first lifeboat station at Formby. During his tenure as Dock Master, William kept detailed accounts of the River Mersey's tides and weather conditions, the earliest continuous record of tidal information in Britain, which enabled him to predict optimal times for shipping to enter or leave the docks. His forward thinking led to the first tide tables being produced, which form the basis of modern tidal recording today.

A channel alongside Perch Rock in New Brighton, which William recommended be deepened to reduce shipwrecks, was named after him in the late 1700s and a

Liverpool Docks with the Liverpool Cathedral (Liverpool Anglican Cathedral, St James's Mount) top right.

*Right*: Thomas Steers' 'Old Dock', now under the Liverpool One shopping complex.

*Below*: From right to left, the Royal Albert Dock, the Three Graces and the Liverpool waterfront, the northern dock network and Seaforth Docks.

*Above*: Liverpool Docks with the Maritime Museum on the right and Royal Albert Dock Pump House in the centre.

*Left*: Liverpool Canning Docks with paddle boarders (centre).

The beautifully restored *Brocklebank* tug.

commemorative fountain dedicated to him was created at Liverpool ONE shopping centre in 2008.

The Royal Albert Dock has become one of the most famous locations in Liverpool and a tourist hotspot. The 'Royal' was awarded in 2018 when it was granted a royal charter, 172 years after the dock was opened in 1846. The dock was designed by Jesse Hartley and Philip Hardwick and was the first to use hydraulic cranes, powered by the nearby pump house; the cranes were employed to quickly and efficiently unload cargo directly from ships into the network of surrounding warehouses. The security features of the dock made it ideal for storing high-value cargo such as tobacco, spirits, sugar, ivory and silk. The dock was revolutionary in design but relatively small, unable to accommodate larger shipping, and so by 1972 the dock had closed. Following a major investment and redevelopment programme, the dock was officially reopened by HRH the Prince of Wales in 1988 and today houses Tate Liverpool, the Beatles Story exhibition and the Merseyside Maritime Museum.

Housed within the Merseyside Maritime Museum is the International Slavery Museum, a permanent exhibition encompassing three galleries which detail the transatlantic slave trade, the history of slavery and life in West Africa, with individual first-hand accounts and historical documentation. Free to enter, the museum should be on everyone's list of things to see in Liverpool.

*Above left*: The Beatles Story.

*Above right*: Royal Albert Dock.

*Opposite*: Mersey Ferry at Seacombe.

*Below*: View across the Canning Docks towards the RIBA North (National Architecture Centre) on the left and Home England Office in the centre.

MERSEY FERRIES SEACOMBE

The best way to arrive in Liverpool is, of course, via the river. To fully appreciate the grandeur of the waterfront and the iconic buildings that overshadow the port and surrounding docks, visitors should take the ferry from Birkenhead and marvel at the skyline that greets you across the Mersey. The Royal Liver Building, the Cunard Building and the Port of Liverpool Building are collectively known as the Three Graces and dominate the Pier Head.

The Three Graces are located just beyond the Museum of Liverpool and behind the Pier Head Ferry Terminal. The Port of Liverpool Building, formally the Mersey Docks and Harbour Board Building, was completed in 1907 following a design competition, won by Arnold Thornely and F. B. Hobbs in collaboration with Briggs and Wolstenholme. Built in an Edwardian Baroque style, clad in gleaming white Portland stone and topped with a most impressive central dome, the building looks more like a palace than an administrative headquarters.

Alongside the Port of Liverpool Building sits the Cunard Building, designed by William Edward Willink and Philip Coldwell Thicknesse, completed in 1917 to be the new headquarters of the Cunard Steamship Company.

*Above left*: Merseyside Maritime Museum.

*Above right*: View from the top of the Liver Building with the Museum of Liverpool and Royal Albert Dock behind.

Liverpool Pier Head.

Perhaps the most famous of the Three Graces—and certainly a building that has become synonymous with Liverpool—is the quintessential Royal Liver Building. Designed by Walter Aubrey Thomas, it was completed in 1911 for the Royal Liver Assurance Group. Standing at 340 feet (103 metres) tall, the building was a skyscraper in its day and the tallest building in Britain when it opened. The building is crowned by two 'Liver Birds', based on cormorants and designed by Carl Bernard Bartels, a German sculptor who won the competition to design the iconic figures. Named Bella and Bertie, myth has it that Liverpool would no longer exist if the birds were to ever fly away.

Home to the Mersey Tunnel Police, Merseytravel and the Mersey Tunnels staff is the splendid George's Dock Building. Sitting behind the Port of Liverpool Building and overshadowed by the Three Graces, the George's Dock Building was designed by Herbert James Rowse in 1934 and houses the ventilation machinery for the Mersey Tunnels. Six ventilation towers were completed in total, three on each side of the river and all designed with magnificent art deco details. Two sculptures sit on either side of the entrance to the George's Dock Building carved in black basalt, representing 'day' and 'night', and one of the original emerald-green tollbooths adorns the terrace at the north of the building.

*Above left and above right*: The Liver Building.

*Left*: One of the famous Liver Birds—Bertie, looking inland over the families of sailors out at sea.

*Opposite above*: The Liverpool skyline. George's Dock Building bottom right.

*Opposite below*: Mersey Tunnels control room.

SIDNEY STREET

*Left*: George's Dock Building with original emerald green tollbooth and lamppost at the foot.

*Below*: Liverpool James Street Station—Mersey Tunnel.

Flowing through Pier Head, the Liverpool Link Canal connects the remaining southern docks to the northern dock network and via locks at Stanley Dock, to the Leeds-to-Liverpool Canal. Opened in 2009, the canal flows in front of the Three Graces, the location of the former George's Dock, remnants of which can still be found under the Royal Liver Building and George's Dock Building.

Mersey Railway Tunnel connects Liverpool (James Street Station) to Birkenhead (Hamilton Square) and was the first tunnel dug under the Mersey and the world's second oldest underground railway, the first being the London Underground. Initially, steam engines were used to carry passengers beneath the river but despite a ventilation system, the tunnels quickly became polluted with smoke and customers returned to the ferries. By 1900 the railway company was bankrupt and was only saved when additional funds were raised and the line was electrified in 1903.

If you fancy a train journey below the waters of the Mersey, take a historic route down to James Street Station via the Water Street underground pedestrian passage. Liverpool James Street opened in 1868 as the Liverpool Mersey Tunnel terminus but was later connected to Liverpool Central Station in 1892. The Water Street passage takes pedestrians under the India Building, beneath Brunswick Street and into the underground station itself. The passage still features the unique design and decor of the station with many original features.

*Below left*: Water Street passageway to James Street Station with some original nineteenth-century detail.

*Below right*: The Water Street entrance to the underground passage leading to James Street Station.

*Above*: The Tobacco Warehouse.

*Below*: The Regent Road Bascule Bridge.

Beyond the Cruise Ship Terminal, the remaining northerly docks dominate the riverside and pedestrians must retreat to Waterloo Road which runs adjacent to the Mersey. The road runs for about half a mile (0.7 kilometres) before transforming into Regent Road, and a little further on crosses the waterway linking Stanley Dock to Collingwood Dock via the historic Regent Road Bascule Bridge, built in 1932.

Jesse Hartley designed both the Stanley and Collingwood Docks that were completed in 1848. The Stanley Dock is flanked by two massive five-storey warehouses; the northern side, commonly known as the rum warehouse, was redeveloped in 2010, becoming the Liverpool Titanic Hotel. The southern warehouse remains but now sits behind the Stanley Dock Tobacco Warehouse, which was built between 1897 and 1901 by Anthony George Lyster and claimed to be the largest building in the world when it opened; Lyster split and backfilled the original dock to build the gigantic fourteen-storey warehouse in its place. The Tobacco Warehouse is nearing the completion of an ambitious redevelopment programme, converting the warehouse into 550 luxury apartments.

Another building recently completed is the new Everton Stadium, constructed on the backfilled Bramley-Moore Dock, replacing Goodison Park as the football team's home ground and set to host matches in the UEFA European Championship in 2028.

Of more interest to me is a Gothic Revival clock tower that stands between the two river gates of Salisbury Dock. Built in 1848 by Jesse Hartley and based on a sketch by

Stanley Dock from Collingwood Dock.

*Above*: Victoria Clock Tower between the river locks of Salisbury Dock.

*Left*: Two canal barges crossing Collingwood Dock before entering Salisbury Dock with the Victoria Clock Tower in the background.

railway architect Philip Hardwick, the Victoria Clock Tower stands at 323 feet (98.5 metres) tall—slightly taller than the Elizabeth Tower on the Palace of Westminster in London. The hexagonal tower's six clocks allowed shipping out on the river as well as in the dock to keep accurate time. In addition, the tower housed a warning bell and provided the pier master with lodgings. Currently in a very poor state, the tower is part of a regeneration programme that should see the surrounding docks restored and repurposed as community spaces and housing. The modern development of the city's waterfront was responsible for Liverpool losing its World Heritage Status in 2021 due to 'irreversible loss' to the heritage value of the Victorian docks.

The Port of Liverpool runs for 7.5 miles (12 kilometres) along the Liverpool side of the Mersey from Brunswick Dock to the most northerly Seaforth Dock, which opened an extension called Liverpool2 in 2016 to handle the largest container ships. The massive dock and container terminal is notable for the sixteen CRMG (cantilever rail-mounted gantry) cranes that dominate the skyline and have become local landmarks, clearly visible from New Brighton and even 10 miles down the Mersey at Eastham.

The beach at Crosby provides a suitable location to end our journey along the River Mersey, with views across the river to New Brighton and the towering cranes at Seaforth, a reminder of the Mersey's significance to British trade and industry. You are never alone on Crosby beach as Antony Gormley's 100 cast-iron figures share the view and witness the river dilute and dissolve into the Irish Sea.

But the river's journey is endless, reborn again and again as somewhere high among the Pennines or across the wild moorlands of Derbyshire, a tiny stream forms among the marshes and heather. Gravity and the landscape takes the stream on a voyage, down through valleys and woodland, gathering pace on its infinite journey to the sea.

Victoria Clock Tower with the new Everton Stadium to the left.

*Above and below*: Seaforth Docks.

Epilogue

# Home

Home is about an hour's drive from Crosby Beach but I prefer to take my time and stop at the waterfront in Liverpool.

I feel an affinity with Liverpool—which may sound odd coming from a Mancunian, but I've always felt there's far more that unites us than divides us, the River Mersey being one of them. From the 1730s, the Mersey and Irwell Navigation connected the cities and shipping could transport goods from Water Street Quays in Manchester directly to Liverpool and the Irish Sea. The world's first inter-city railway linked Manchester and Liverpool, designed and built by George Stevenson in 1830, and was so successful that it influenced the development of railways across the world.

A lone figure looks out to sea—one of Antony Gormley's cast-iron figures, part of the 'Another Place' installation.

Politically and economically, the cities have a lot in common and despite the odd squabble over football, I hold a brotherly affection for my neighbouring scousers.

The affection I feel for the river I grew up less than half a mile from is in no doubt, and the attachment has only grown stronger with time. Rivers hold a profound spiritual meaning in most cultures, a symbol for renewal, rebirth, purification ... the beginning or end of a journey!

From Liverpool I like to drive back towards Widnes and cross the Mersey via the Jubilee Bridge and meander along the A588 and A56 towards Warrington and join the M56 near Lymm—the river isn't visible but I sense its proximity. I exit the M56 at junction 1 and take the A34 towards Manchester, almost immediately crossing the River Mersey via the Didsbury Kingsway Bridge. I can glimpse the river as it passes beneath the neighbouring railway bridge and I know I am home.

Home.